Tea Plantation Management

THE AUTHOR

Dr. S.E. Kabir is the Director of Siliguri Tea Training Institute. He was the Assistant Professor in the Department of Tea Science, University of North Bengal. He was associated with Tea Board India for Tea Research works. He also works as Technical Expert for Biotechnology Industry Research Assistance Council (BIRAC) and Agriculture Skill Council of India (ASCI)

Tea Plantation Management

Dr. S. E. Kabir

Director

Siliguri Tea Training Institute

2025

Regency Publications

A Division of

Astral International Pvt. Ltd.

New Delhi – 110 002

ISBN: 978-93-5919-626-8

Published by : **Regency Publications**
A Division of
Astral International Pvt. Ltd.
– ISO 9001:2015 Certified Company –
4736/23, Ansari Road, Darya Ganj
New Delhi-110 002
Ph. 011-43549197, 23278134
E-mail: info@astralint.com
Website: www.astralint.com

PREFACE

There are many very good books published on tea cultivation. This book is an addition to those books, with detailed features

The content of this book may not conform to the findings of the Tea Research Centres or other organizations. It is not at all obligatory on the part of the tea gardens / tea planters or any body else to follow the instructions and advices provided in this book.

Many pictures and descriptions have been taken from various books, which have been duly acknowledged. Publications of Tocklai Experimental Station, Jorhat and UPASI Tea Research Foundation, Valparai, which are the most renowned Tea Research Centres in the world are the main strength of this book. The authors cordially and humbly acknowledge their support in bringing out this book.

The author is indebted to Tea Board India, Tocklai Tea Research Institute, Dr. S.K.Debnath, Tea Scientist , Md. Fareed, Former Senior Advisory Officer & Principal Scientist, Tea Research Association for their immense contribution to this book. The authors express deep sense of gratitude to S.B.Agency, Siliguri for providing many data on pesticides etc.

This book is supposed to serve as a tool for the tea planters, especially the new comers in the field of tea and the students pursuing tea courses. The authors have tried to make the book as simple as possible and avoid too much of technicality. It can serve as a practical hand book for the tea planters.

(Dr. S. E. Kabir)

CONTENTS

Preface *v*

1. **Botany of Tea & Tea Varieties** 1

Conditions Required for Growing Tea; Systematic Position of Tea; Tea Varieties

2. **Soil Science & Nutrients** 3

Components of Soil; Difference Between Top Soil and Sub Soil; Physical Properties of Soil; Factors Affecting Soil Structure; USDA Taxonomy of Soil; Role of Physical Properties in Soil Fertility; Chemical Properties; ION Exchange; Role of Various Nutrients; Available Phosphorus; Deficiency Symptoms; Deduction for Vacancies and Drains/ Road ETC

3. **Vegetative Propagation** 27

Nursery Beds; Sleeve Size; Time of Planting Cuttings; Shade Tree Nursery

4. **Young Tea Management** 32

For Removal of Existing Old Bushes; Rehabilitation;Single Hedge Plantation; Double Hedge Plantation; Staking; Pit Size; Inputs in The Planting Pits; Young Tea Manuring; Post FFP Treatments

5. **Pruning** 42

Time of Pruning; How to Decide Pruning Time ?; Starch Test; Mature Tea Pruning

6. **Plucking** 47

Periodicity of Growth; Interflush Dormancy; Causes of Interflush dormancy; Causes of winter dormancy; Plucking; Tipping; Mechanical Harvesting; References

7. **Drainage & Irrigation** 55

Benefits of Drainage; Problems Associated with Waterlogging; Few Tips on Irrigation

8. **Shade Trees** 65

Characteristic Features Of Few Major Shade Trees

9. **Bio Organic Farming / Biological Control of Pests** 69

Demerits of Synthetic Pesticides; Biological Pest Control Protocols in Complete Bio Organic Sections; Control of Caterpillars; Calendar of Methodology for A Bio-Organic Tea Garden; Vermiculture

10. **Tea Pests and Pest Control** 91

Definition of Pests; Insects; Root Damaging Pests; Cockchafer Grub; Control Measures; Leaf Eating Pests; Control Measures; Control of Mites; Description of A Few Stem Diseases; A Few Leaf Diseases; Control of Primary Root Diseases; Plant Protection Code : Version 15 : May 2023

11. **Weeds & Weed Control** 114

Harmful Effect of Weeds; Types of Weeds; Manual / Mechanical Control; Characteristic Features of A Few Major Weeds

Colour Plates I-XII

Exercise No. 1

Botany of Tea & Tea Varieties

CONDITIONS REQUIRED FOR GROWING TEA

Locality	Nearer to marketing place, transport and trading facility
Depth of Soil	Pervious up to 5-6 feet
Physical Properties of soil	Deep and porous
Drainage	Good drainage with sufficient out fall
pH of Soil	4.5 to 5.5
Ambient Temp	20 – 30 C (maximum 35 C)
Leaf Temperature	Min 21 C (maximum 39 C)
Soil Temperature	Not less than 20 C; Optimum 25 C
Day length	Not less than 11 hours 15 minutes
Rainfall	1000 – 1400 mm of annual rainfall; 100-150 per month on an average
Relative Humidity	80-90 %

SYSTEMATIC POSITION OF TEA

Order : Theales

Family : Theaceae

Genus : *Camellia*

Species : *sinensis*

Botanically, tea is a tree. Tea tree is modified to a shrub by pruning with the following objectives :

i. To keep it under pluckable height.
ii. To induce lateral growth for producing more plucking points.
iii. To keep the plant in vegetative form because the commercial product of tea comes from the leaves of the tea bushes and not fruit or seed.
iv. To encourage more flush, more productivity.

TEA VARIETIES

There are mainly three varieties of tea :

1. China Variety - *Camellia sinensis* (L) O.Kuntze.

It is a shrub with multistemming character. All the stems emerge from the soil directly. There is no distinct trunk. Leaves of China variety tea is very small and very high in Volatile Flavour Constituents (VFC) which imparts 'Muscatel Flavour'. Flavour Index is the ratio of VFC II and VFC I. In VFC IIs, Geraniol is the main component which emits rosy flavour . Apart from geraniol, it contains Linalool and its oxides. The leaves have orthotropic (erect) growth habit. Length of the leaves varies from 1.5 to 14 cm and width 1 to 2.5 cm. China variety is known for its flavour.

2. Assam Variety – *Camellia assamica* (Masters) Wight.

It is bigger in size than China variety . Shrub is vigorous in growth. Leaves are bigger than that of China variety. Length varies between 8 to 20 cm and width varies between 3 to 8 cm. The VFC content in Assam variety is less than that of China variety. Assam variety is known for its dark cup character with high concentration of polyphenols.

3. Cambod Variety – *Camellia assamica ssp lasiocalyx* (Planchon ex Watt) Wight.

It has the features like Assam variety but has broadly elliptic (boat shaped) leaves with not too distinct marginal veins, often with pink anthocyanin pigmentation on the base of petiole particularly when young. Leaves are having orthotropic growth habit and pale green colour.

Exercise No. 2

Soil Science & Nutrients

Soil is defined as "The uppermost loose layer of the earth suitable for plant growth."

COMPONENTS OF SOIL

Mineral Matter (45%)

Organic Matter (5%)

Soil Water (25%)

Soil Air (25%)

DIFFERENCE BETWEEN TOP SOIL AND SUB SOIL

Top Soil (0-22.5 cm from top surface)	Sub Soil (below 22.5 cm depth)
Completely weathered	Not completely weathered
Soil particles are fine	Particles are coarse
Aeration is more	Poor aeration
More number of micro organism	Number is less
Higher organic matter content	Lower organic matter content
It is dark in colour	It is light in colour
High in Nitrogen content	High in potassium content
It is fertile	Sub soil is less fertile

PHYSICAL PROPERTIES OF SOIL

Physical properties of a soil influences plant growth. Following are the important physical properties of soils:

Soil texture, Soil structure, Soil density, Soil porosity, Soil consistency, Soil colour, Soil temperature.

SOIL SEPARATES

According to the size, the soil particles are grouped into gravels, sands, silts and clay which are termed as Soil Separates.

Soil Separates	Diameter Range(mm)
Coarse Sand	2.00 – 0.20
Fine Sand	0.20 – 0.02
Silt	0.02 – 0.002
Clay	Below 0.002

TEXTURAL CLASS

Textural group	Sand %	Silt %	Clay %
Sand	80-100	0-20	0-20
Sandy loam	50-80	0-50	0-20
Loam	30-50	30-50	0-20
Silt loam	0-50	50-100	0-20
Sandy clay loam	50-80	0-30	20-30
Silty clay loam	0-30	50-80	20-30
Clay loam	20-50	20-50	20-30
Sandy Clay	50-70	0-20	30-50
Silty Clay	0-20	50-70	30-50
Clay	0-50	0-50	30-100

SOIL STRUCTURE

There are four principal forms of structure:

1. Plate-like – This type of soil structure is mostly noticeable in the surface layers of virgin soils but may be present in the sub soil.
2. Prism-like – This type of soil structure is commonly occurring in sub soil horizons.
3. Block – like - This type of soil structure is mostly noticeable in the sub soil horizons and characteristics have much to do with soil drainage, aeration and root penetration.
4. Sphere- like – All rounded aggregates may be placed in this category. These rounded complexes usually lie loosely and separately.

FACTORS AFFECTING SOIL STRUCTURE

The development of structure in soil depends on the following factors:

1. Climate

2. Organic matter – It improves the structure of a sandy soil as well as of a clay soil.
3. Tillage
4. Plant Root
5. Soil organism
6. Fertilizers – Fertilizer like Sodium Nitrate destroys granulation by reducing the stability of aggregates. Few fertilizers, for examples, Calcium Ammonium Nitrate help in development of good structure.

SOIL DENSITY

Particle density - The weight per unit volume of the solid portion of soil is called particle density. Particle density of a normal soil is 2.65 g/ c.c.

Bulk Density – The oven dry weight of a unit volume of soil inclusive of pore spaces is called bulk density. The bulk density of soil is always smaller than its particle density. The bulk density of sandy soil is about 1.6 g/c.c.

POROSITY OF SOIL

Soil porosity is the percentage pore space. It can be calculated by the formula:

Porosity (% pore space) = 100 – (bulk density / particle density) X 100

PORE SPACE

The volume of soil mass that is not occupied by soil particles is known as pore space. The pore space is usually occupied by air and water. In the pore spaces, the plant roots grow and exit. It directly controls the amounts of water and air in the soil and thus indirectly controls plant growth and crop production. In general two sizes of pores are recognized i.e Macro or non- capillary and Micro or capillary.

SOIL COLOUR & COMPOSITION

a. Black or dark grey: The variations from black to dark grey colour of soil are mainly due to organic matter.
b. Brown colour: This is the most common soil colour and is due to a mixture of the organic matter and iron oxides.
c. Red – yellow colour: Red colour is associated with anhydrate ferric oxides, whereas yellow colour indicates some degree of hydration.
d. White colour: Sillica and lime generally impart white colour.
e. Bluish & Greenish colour: Due to the presence of ferrous compounds in ill- drained soil.

IMPORTANCE OF SOIL TEMPERATURE

Effect of soil temperature on soil microorganism: Soil temperature controls the rate of microbial process involved in mineralization of nitrogen, composition of organic matter, nitrogen-fixation etc. There is lowest microbial activity where soil temperature is below 5°C and above 54°C.Optimum temperature is between (25 – 35)°C.

Effect of soil temperature on the decomposition of soil – organic matter: The higher the temperature the more rapid is the decomposition.

Effect of soil temperature on absorption of soil water: Very low and very high temperature affects the absorption of soil water by the roots of plants.

Effect of soil temperature on nutrients availability: The solubility of food nutrients and their availability are considerably influenced by soil temp.

Effect of soil temperature on seed germination: If the temp. is too low, the seed fails to germinate or germinate very slowly. If it is too high, seed may be injured.

Effect of soil temperature on plant growth: During plant growth physiological processes are influenced by temperature.

Effect of soil temperature on soil formation: Process of weathering and soil development are accelerated where there is an increase in soil temp.

Effect of soil temperature on plant diseases: Development of diseases is also related with soil temperature.

SOIL DEPTH

Tea grows best in deep soil. A soil can be called deep if it is 90cm or more in depth, it is moderately deep when the depth is 50-90cm and only shallow when the depth is 25-50cm.

Different management practices are used for varying depths of soils under mature tea, young tea and for replanting. If very shallow (0-25cm), or merely shallow

(25-50cm) depth occurs in soils under mature tea area then the following steps are to be taken:

a. Maintenance of good shade
b. Split application of fertilizer
c. Irrigation at frequent intervals with lower amount
d. Shorter pruning cycle
e. Chemical weed control

SOIL PROFILE

It is a vertical section of the soil through all its horizons and it extends up to the parent material. With the increase in depth changes in texture and structure of the soil is noticed. The soil column in most cases is composed of a series of layers which are parallel with the soil surface and are called soil horizons.

Oi – Organic Slightly decomposed

Oe – Organic Moderately decomposed

Oa – Organic Highly decomposed

A – Mineral. Mixed with humus, dark coloured

E – Horizon of maximum accumulation of silicate clays, Fe and Al oxides

AB or EB – Transition to B, More likely A or E than B

BA or BE – Transition to A or E, more likely B than A or E

B – Most clearly expressed portion of B horizon

BC – Transition to C, more likely B than C

C – Zone of least withering, accumulation of Ca, Mg carbonates , cementation, sometimes high bulk density

R – Bed Rock

USDA TAXONOMY OF SOIL

SURFACE HORIZON

Mollic (A):	Thick, dark coloured, high base saturation, strong structure.
Umbric (A):	Same as Mollic except low base saturation
Ochric (A):	Light coloured, low organic content
Histic(O):	Very high in organic content
Anthropic (A):	Man modified Mollic like, high in available Phosphorus
Plaggen (A):	Created by year of manuring.

SUB-SURFACE HORIZONS

Argillic (Bt) : Silicate clay accumulation.

Natric (Btn) : Argillic , high in Sodium, columnar or prismatic structure.

Spodic (Bhs) : Organic matter, Fe and Al Oxide accumulated.

Cambic (B) : Changed or altered by physical movement or by chemical reactions.

Agric (A or B) : Organic or altered by physical movement or by chemical reactions.

Oxic (BO) : Highly weathered, mixture of Fe, Al oxides.

Duripan (M) : Hard pan, strongly cemented by silica.

Fragipan (X) : Brittle pan, loamy textured, weekly cemented.

Albic (E) : Light coloured , clay and Fe, Al oxides mostly removed.

Calcic (K) : Accumulation of $CaCO_3$ or $CaCO_3$. $MgCO_3$.

Gypsic (Y) : Accumulation of gypsum.

Salic (Z) : Accumulation of Salts.

ROLE OF PHYSICAL PROPERTIES IN SOIL FERTILITY

Importance of Texture: Sandy soils have poor fertility and very little water holding capacity so only few crops can be grown such as potato, groundnuts etc. Clay soils have high fertility and very high water holding capacity. Rice, jute, sugarcane can be grown in this soil. Silt (loam) soils have medium fertility and water holding capacity. They are suited to every kind of crop such as wheat, maize, pulses, vegetables, sugarcane etc.

Importance of Structure: Platy structure normally results in water logging. This structure is not good for cultivation .Whereas sphere – like structure helps in drainage and provides optimum aeration, water holding capacity etc. This structure is considered best for cultivation.

Importance of Density: Soils with low bulk density have favourable physical conditions.

Importance of Porosity: For ideal conditions of aeration, permeability and water holding capacity, a soil should have an equal amount of macro and micro pores.

Importance of Consistence: Optimum soil consistence increases the water holding capacity and plant food material. Thus, increasing the fertility of soil.

Importance of Soil Colour: Soil colour are indirectly related with soil fertility:

a. Relation of soil colour with temperature: Directly the soil colour influences soil temperature to some extent. The dark colored soils absorb more heat than light colored soil.

b. Relation of soil colour with organic matter: organic matter imparts black to grey tinges. More quantity of organic matter content improves the soil structure.

c. Relation of soil colour with drainage: Bluish and greenish colour of soil indicate water logging. Need of drainage can be determined by the colour variations in the soil.

d. Relation of soil colour with parent material: soil colour may indicate their origin from parent material.

Note: The information on Soil Science have been derived from the books i) Fundamentals of Soil by V.N.Sahai ii) A Text Book of Soil Analysis- T.C.Baruah & H.P.Borthakur)

INFLUENCE OF SOIL AND OTHER PARAMETERS ON SOIL APPLIED FERTILIZER USE EFFICIENCY

1. SOIL TEXTURE: Lighter the soil texture i.e. higher the sand content greater is the porosity but lesser is the retention of nutrients (N, K, Ca, Mg, and Al) and water.

SUGGESTIONS

a. Enrich organic matter as much as possible by extraneous application.

b. Stress on more foliar application as long as the soil is enriched with desired organic matter content.

c. Avoid application of more organic matter in soil having high silt content. This will reduce the porosity of soil.

2. SOIL STRUCTURE : Generally poor in low land under cultivation of paddy and planted in non-rehabilitated area

SUGGESTIONS

a. Light forking in Light Pruned years

b. Apply dolomite as a source of calcium

c. Get rid of water logging by good drainage

d. Enrich organic matter content

e. Check the root distribution as a sample (uproot and check). If it is poor, give more stress on foliar application till the root develops properly.

3. ORGANIC MATTER CONTENT – Higher the organic matter content better is the fertilizer use efficiency. Target should be to get a organic carbon more than 2 %. Lesser the organic matter content, higher the loss of fertilizer through leaching, lesser the sustained available nutrient content, higher the requirement of fertilizer to compensate the loss.

SUGGESTIONS

a. Add any source of organic matter (cattle manure, oil cakes, decomposed tea waste, decomposed shrubs, weeds, vermicasting etc)

b. Grow Guatemala grass / Citronella in vacant patches until it is in filled.

4. SOIL pH

pH: Below 4.5 – Not suitable for tea cultivation

4.5 to 5.5 – Suitable

Above 5.5 – Not suitable

SUGGESTIONS

a. If pH is high, adjust with aluminium sulphate or pyrites

b. If pH is low, adjust with dolomite

5. FERTILITY STATUS : Soil fertility depends on nutrient content of the soil and soil micro organisms.

SUGGESTIONS:

a. Every tea estate should have soil report of every section.
b. Soil testing should be done during every LP yea
c. Addition of organic manure along with extraneous application of microbes like *Azotobacter, Azospirillum, Bacillus megatherium* etc in the pruned years will help improvement of soil fertility.

6. DRAINAGE STATUS: Improved drainage system can assure higher fertilizer use efficiency.

SUGGESTIONS

a. Improved drainage condition is the top priority
b. Till drainage is improved, foliar application of urea and potassium plays a great role.
c. If soil phosphate status is low and phosphorus is bound due to drainage problem, spray soluble phosphate (DAP) instead of applying ground phosphate.

CHEMICAL PROPERTIES

1. Soil pH
2. Ion Exchange
3. Available Nitrogen
4. Available Phosphorus
5. Available Potassium
6. Available Calcium
7. Available Sulphur
8. Available Magnesium
9. Available Zinc
10. Available Boron

SOIL pH

Soil pH is defined as the negative logarithm of hydrogen ion concentration. It indicates the acidity or alkalinity of soil. pH of water is 7 which is neutral. pH range of soil may usually vary from 3.5 to 10.5. Soil pH below 7 indicates soil is acidic and below 7 indicates soil is alkaline. Tea grows well in pH between 4.5 to 5.5.

A change in of one unit of pH indicates 10 times increase or decrease in hydrogen ion concentration.

Reasons of low pH are:

- Too much use of urea, even in foliar spray it is used.
- Garden near to dolomite mine
- Inadequate drainage
- If pH is low (below 4.5) it can be increased by applying dolomite @ 2 tons/ha in Nov-Dec in coarse textured soil and @ 3 tons/ha in fine textured soil.
- If pH is high (between 5.6-5.8) in can be lower by applying Aluminum Sulphate or Sulphur Pyrites.
- Aluminum Sulphate is more costly but where organic matter is less there Aluminum Sulphate is used instead of being costly because Pyrites need microbial activity to release.

ION EXCHANGE

Ions may be positively charged and negatively charged. Ion exchange is a reversible process by which cations and anions are exchanged between solid and liquid phases. The exchange of cations and anions are known as cation exchange and anion exchange respectively.

ROLE OF VARIOUS NUTRIENTS

AVAILABLE NITROGEN

Nitrogen in soil existed mostly in the form of compounds resulting from decaying vegetable matter and was present in complex and simple organic compounds, as well as in the form of ammonia, nitrites and nitrates.

The major source of nitrogen in soil is organic matter.

Generally the nitrogen content progressively depleted over the period of cropping. In Jorhat area, the depletion between virgin jungle soil and old tea soil was more pronounced after forty years of cropping.

Application of the chemical significantly influences the soil nitrogen, carbon, and C:N ratio. A higher dose of nitrogen (135kg/ha) resulted in better nitrogen, carbon and C:N status of the soil than a lower dose of nitrogen (45 kg/ha)

Total nitrogen of the top soil was found highly crop related with yield. Pruning litter and Nitrogen fertilizer application of up to 200 kg/ha helped the residual build up of soil Nitrogen and increased shoot Nitrogen.

Nitrogen is not easily available to the plant directly. There are some nitrogen fixing bacteria like *Nitrobactor, Azotobacter, Nitrosomonus* that fix the atmospheric nitrogen through Nitrogen Fixation.

AVAILABLE PHOSPHORUS

The optimum range of available Phosphorus should be 10 to 20 ppm.

Applied as i) DAP- Foliar Spray
ii) RP- Acid Soluble
iii) SSP- Water Soluble

DAP should be given as 1% foliar spray (Total 22 kg/ ha/ year.)

After huge attack of looper, red spider mite, hail storm should go for spray of DAP.

DAP and Zn are incompatible so DAP should not be applied with Zn as a cocktail..

Blanket Dose: 50 kg P_2O_5 /ha/year

PHOSPHORUS

1. Immobile in soil as well as in leaf.
2. It will not leach out.
3. Rock phosphate is slow releasing, it is persistent in soil releases gradually.
4. RP is not water soluble so during heavy rain also it will not lost.
5. If PH is more than 5.5 then SSP should be applied & if pH is low then RP should be applied.
6. Phosphate has a tendency to bind with Al, Ca, Fe. If once bind then it is not available to the plant.
7. *Bacillus megatherium* and VAM activates phosphorus uptake of the plant.
8. Phosphate is required 30 g/plant.
9. Placement method is better in case of phosphorus.
10. In Feb-March along with organic manure 30 g of RP is given per pit and it is placed between 2 plants.
11. It helps in rooting of the plant.

POTASSIUM

1. Potash controls the stomatal opening and give strength to the plant.
2. Due to potash deficiency plant become weak, short, weepy, thin, top surface becomes breakable, full of die back will be there.
3. Potash is given as Muriate of Potash. MOP is a luxury nutrient
4. because, a) it has no toxicity

5. Doesn't require microbial action for activation
6. In higher dose also it will not affect the plant

As per latest finding, when lower pruning is done potash requirement increases.

If less than 45cm, N:K_2 O :: 1:2 (In case of RP,CP)

If 45 cm to 60 cm, N:K_2 O :: 2:3

If above 60 cm, N:K_2 O :: 1:1

During drought condition when stomatal conductance increases then foliar application of potash should be given starting from December to April (fortnight).

During rainy season when the area is submersed then foliar application of potash should be given.

SULPHUR

Sulphur is an essential plant nutrient having three amino acids (cystine, cystiene and methionine) which serve as the precursor of VFC synthesis.

Optimum range of Sulphur in soil is more than 30ppm.

It is applied as elemental sulphur, @ 50 kg in 2 splits ; Pre & Post monsoon.

Sulphur should not be applied during rainy season. Because it reacts with water and form H_2S which has a toxic effect to the plant.

ZINC

Zinc is an essential plant nutrient absorbed by roots as Zn^{++}.

About 62% of soils studied had Zn between 3ppm-15ppm and another 29% between 15ppm-27ppm. Only 4% of the soils had Zn above or below the limits of 3 ppm to 27 ppm.

Available Zinc of soils did not vary much vertically up to a depth of 120cm.

About 4.6% of the total zinc remained in available form in tea soils, and both total and available zinc were linearly correlated (r2 = 0.66). The total Zn content of tea soils vary from 170 ppm to 570 ppm.

BORON

It is one of the essential micronutrients absorbed by roots as $H_2BO_3^-$ or $H_3BO_3^{-}$

Total Boron content varied from 8.80ppm to 25 ppm, and available Boron from 0.88 ppm to 0.44 ppm. Available Boron content of tea soils is low being 1% of the total in the South bank, and between 1.5% to 1.8% in the North bank.

Available Boron has a strong positive correlation with total Boron quantity.

Soil availability of Boron decreases significantly with increasing level of Ca application.

ROLE OF NITROGEN

1. Nitrogen is a part of all living cells and is a necessary part of all proteins, enzymes and metabolic process involved in the synthesis and transfer of energy. Affect energy reactions in the plants.
2. Essential components chlorophyll, which is responsible for photosynthesis. Imparts dark green colour to tea plants.
3. Present in nucleotides, phosphatides, alkaloids, enzymes, hormones, vitamins etc.
4. Necessary for formation of amino acids, the building blocks of protein.
5. Promotes leaf, stem and other vegetative growth.
6. Aids in production and use of carbohydrates.

ROLE OF PHOSPHORUS

1. Phosphorus is an essential part of the process of photosynthesis.
2. Involved in the formation of all oils, sugars, starches etc.
3. Constituents of nucleic acids, phytin, and phospholipids.
4. Encourages blooming and root growth.
5. Promotes early root formation and growth.
6. Responsible for cell division and enlargement.
7. Helps with the transformation of solar energy into chemical energy; plants survive harsh winter conditions.
8. Increase water-use efficiency and hastens maturity.

ROLE OF POTASSIUM

1. Vital to photosynthesis, essential to protein synthesis.
2. Increases water use efficiency and turgidity; regulates stomatal function
3. Increases Nitrogen metabolism, and breakdown of carbohydrate & translocation of carbohydrate from leaf to storage organs; translocation of starch.
4. Activates enzymes and controls their reaction rate.
5. Counteracts the injurious effect of excess Nitrogen.
6. Imparts increased vigour and disease resistance to plants.
7. Improves the quality of seeds and fruit.
8. Improves winter hardiness.

ROLE OF MAGNESIUM

1. Key element of chlorophyll, proteins, enzymes, and certain volatile compounds.
2. Very much essential for photosynthesis.
3. Helps in synthesis of carbohydrates, protein, fat and vitamins.
4. Regulates uptake of other nutrients.
5. Increase iron utilization in plants.
6. Activator and component of many plant enzymes.
7. Improves utilization and mobility of phosphorus.

ROLE OF CALCIUM

1. Constituent of cell wall structure, provide strength to the plant.
2. Promotes early root development and growth.
3. Neutralises poisonous organic acids and alkali salts.
4. Activates root tips.
5. Improves uptake of other plant nutrients.
6. Involved in nitrogen metabolism.
7. Reduces plant respiration.
8. Stimulates microbial activity.

ROLE OF SULPHUR

1. Constituents of Sulphur bearing amino acids like cystine, cystiene and methionine.
2. Acts as a precursor of Volatile Flavour Constituents.
3. Promotes nodule formation.
4. Essential plant food for production of protein.
5. Promotes activity and development of enzymes and vitamins.
6. Helps in chlorophyll formation.
7. Helps with vigorous plant growth and resistance to cold.

ROLE OF ZINC

1. Constituents of several enzyme system, which include alcoholdehydrogenase, lactic acid dehydrogenase, glutamic acid dehydrgenase, carboxypeptidase.
2. Influences the formation of growth hormones in plants.
3. Zinc is required in the synthesis of tryptophane, a precursor of Indole Acetic Acid. Formation of this auxin is influenced by Zinc.
4. Zinc has a possible role in plant metabolism involved in starch formation.

5. Associated with water uptake and water relations in plants.
6. Regulate composition of sugar.
7. Necessary for chlorophyll production.
8. Necessary for starch formation.

ROLE OF BORON

1. Keeps Calcium in soluble form and increases its mobility.
2. Helps in absorption of Nitrogen and Carbohydrate.
3. Constituent of cell membrane and essential for cell wall formation.
4. Precipitates excess cations.
5. Helps in translocation of sugars.
6. Essential for germination of pollen grains and growth of pollen tubes.

ROLE OF MANGANESE

1. Supports movement of iron in the plant.
2. Helps in chlorophyll formation.
3. Acts as catalyst in oxidation reduction reactions within the plant tissues.
4. Helps in respiration and protein synthesis.
5. Increase the availability of Phosphorus and Calcium.

DEFICIENCY SYMPTOMS

NITROGEN

1. The colour of the flush and young leaves become light green
2. Older leaves turn brown and fall.
3. The leaves under bright sunshine looses colour very rapidly.

PHOSPHORUS

1. The leaves become dull & matty.
2. Shine of the leaf is lost.
3. A bluish tint develops at the later stages.
4. Die back of stems possible.
5. During phosphate deficiency, stems die back from the ends which have been cut for pruning.

POTASSIUM

1. The tip and the apical margins becomes black coloured.
2. Sometimes leaves turn red bronze in colour.

3. On shaking a Potassium deficient bush, many leaves fall off.
4. Maintenance canopy becomes scanty.
5. Brown blight *Colletotrichum camelliae* and grey blight *Pestalozzia theae* are associated with Potash deficiency.
6. The bark becomes silvery white.

MAGNESIUM

1. Deficient leaves develop a characteristic marginal yellowing which leaves a dark green V – shaped area;
2. Old leaves are affected first.
3. Inverted V shaped dark green patches develops on the leaf surface.

SULPHUR

1. Deficient leaves become bright yellow between the veins, which remain green.
2. New leaves become smaller and the inter nodal distance is reduced.
3. The leaves scorch and fall off and stems die back from the tips. The condition is known as tea yellows.

ZINC

1. Rosetting of leaves near nodal region.
2. Sickle shaped leaves.
3. Wavy edge leaf margins often go yellow.
4. Short internodes.

BORON

1. This deficiency is also very rare.
2. The apical bud fails to grow and cork develops at the upper side of the leaf petiole.

MANGANESE

1. The deficiency is very rare, occurring occasionally on soil of very high pH.
2. Leaves turn pale yellow on the edges and develop a mottling of red brown sports.

NUTRIENT CONTENT OF FERTILIZERS

1. NITROGEN : Urea (46 % N); Ammonium Sulphate (16 % N)
2. POTASSIUM : MOP (60 % K_2O)
3. PHOSPHATE : SSP (16 % P_2O_5, 19.6% Ca, 11-12 % S) ; Rock phosphate (20-22 % P_2O_5, 33 % Ca) ; DAP (46 % P_2O_5, 18 % N)

4. SULPHUR : Gypsum (18.6 % S) ; Wettable Sulphur (80% S)
5. ZINC : Zinc Sulphate($ZnSO_4$, $7H_2O$) (22 % Zinc, 11 % S) ; Zinc EDTA (10 % Zn)
6. MAGNESIUM : Magnesium Sulphate ($MgSO_4$, $7H_2O$)(9.4 % Mg, 13 % S)
7. BORON : Borax (Na-tetraborate) (10 % B)(soluble in cold water ; Boric acid (H_3BO_3) (10 % B) (insoluble in cold water)
8. DOLOMITE : 21 % Ca, 12 % Mg
9. PYRITE : Pyrite (FeS_2) (52 % S, 46 % Fe)
10. ALUMINIUM SULPHATE : 22% Al, 26 % S

YTD DOSES RECOMMENDED BY TRA

Year of Planting	N (kg/ha)	YTD (kg/ha)	Method of Application
0 Year	20-40	200-400	Ring in 2-3 splits
+ 1 Year	80-100	800-1000	Ring in 4 splits
+ 2 Year	100-120	1000-1200	Ring in 4 splits
+3 Year	120-140	1200-1400	Ring in 4 splits
+4 Year	140-150	1400-1500	Strip in 2 splits

MATURE TEA MANURING (PLAINS) DOSES RECOMMENDED BY TRA

Yield range (MTKHY) Pruning Cycle average	N (kg/ha)	P_2O_5 (kg/ha)	K_2O (kg/ha)		
			Soil available Potash status		
			Low (Less than 60 ppm)	Medium (60-100 ppm)	Medium (60-100 ppm)
Upto 1500	Upto 90	20	Upto 90	Upto 70	Upto 70
1500-2000	90-110	20-30	90-110	70-80	70-80
2000-2500	110-140	30-50	110-140	80-120	80-120
2500-3000	140-150	50	140-165	120-140	120-140

MATURE TEA (HILLS) DOSES RECOMMENDED BY TRA

Yield range (MTKHY) Pruning Cycle average	N (kg/ha)	P_2O_5 (kg/ha)	K_2O (kg/ha)		
			Soil available Potash status		
			Low (Less than 60 ppm)	Medium (60-100 ppm)	Medium (60-100 ppm)
Upto 1500	Upto 90	20	Upto 90	Upto 70	Upto 70
1500-2000	90-110	20-30	90-110	70-80	70-80
2000-2500	110-140	30-50	110-140	80-120	80-120
2500-3000	140-150	50	140-165	120-140	120-140

Another method of calculation for manuring in mature tea sections is as follows:

Upto 2000 kg yield per hectare per year, the dose of Nitrogen should be 160 kg. For every 100 kg increase, 4 kg should be added. For example , if in a section the yield is 2100 kg /hectare, the dose will be 160 + 4 = 164 kg /hectare.

DEDUCTION FOR VACANCIES AND DRAINS/ ROAD ETC

The vacancy percentage (in case tea bushes have died due to some reason) should be calculated manually and the exact percentage should be deducted from the calculated requirement. Suppose in a section the vacancy percentage is 5 % and the total requirement of Nitrogen in that year is 6500 kg., the quantity will stand as 95% of 6500 = 6175 kg. A school of thought believes that there should be no deduction for vacancies because the fertilizer will go to the soil of the vacant area, which will be of some use to the tea plants in future. Generally, for drains and roads 4 % is deducted.

EXAMPLE

What should be the fertilizer doses in a section having following statistics?

Location : Dooars (West Bengal)

Area : 20.5 hectares

Style of Planting : Staggered Single hedge

Spacing : 105 cm X 60 cm

Vacancy : 4 % (due to death of tea plants)

Organic Carbon : 1.5 %

Potash status : Optimum

Sulphur status : Low

pH – 4.5

Productivity (MTKHY) : 3500 kg (Mean of one pruning cycle)

Pruning / Skiffing status in the year of application : Light Skiff (90cm from the ground level)

Nitrogen requirement :

For 2000 kg – 160 kg

For 1500 kg – 60 kg

Total - 220 kg per hectare

In 100 kg Urea – 46 kg of Nitrogen is available

So, for 220 kg Nitrogen , 220 x 100 / 46 = 478 kg Urea is required per hectare

Total area – 20.5 hectares

Hence, total requirement of Urea for 20.5 hectares = 478 x 20.5 = 9799 kg

Deduction for road/drain 4 % + 4 % for vacancy (due to death of tea plants) = 8 %

After deducting 8 % , total requirement of Urea = 9015 kg

HOW MANY SPLITS TO BE GIVEN IN ONE YEAR ?

It is a one-million dollar question. It has been seen that many planters are confused in this regard but the process is very simple.

We know that in one split, the minimum quantity of Nitrogen which can be applied is 30 kg per hectare and the maximum is 100 kg per hectare. Hence, if by calculating the dose comes to 95 kg per hectare then only one split i.e. basal dose is enough. If the requirement is 267 kg per hectare , then 3 splits will be required.

So in this case , the number of splits for 220 kg Nitrogen per hectare will be 3 (three).

TIME OF THREE SPLITS

In case of three splits, the application time should be as follows :

1St Split : Pre-monsoon , after receiving 5 cm of initial rainfall.

2nd Split: Just before the onset of monsoon.

3rd Split : Post monsoon, before the soil moisture dries up.

SPLIT PERCENTAGE

1st Split – 40 %

2nd Split – 30 %

3rd Split – 30 %

So in this case , the split and the quantum of Urea for 20.5 hectares should be as follows :

1st Split – March-April – 9015 kg of Urea x 40 % = 3606 kg

2nd Split – June 1st week – 9015 kg of Urea x 30 % = 2704 kg

3rd Split- Mid September – 9015 kg of Urea x 30 % = 2704 kg

REQUIREMENT OF POTASSIUM

Since Potash status is alright and the section is under Light skiff, the ratio of N: K_2O should be 1 : 1. So the K_2O requirement in the present case is also 220 kg per hectare.

100 kg of Muriate of Potash (MOP) (Potassium chloride) contains 60 % K_2O. Hence for 220 kg K_2O, the requirement of MOP is 220 x 100 / 60 = 367 kg per hectare. For 20.5 hectares total requirement is 7523 kg.

Deduction for road/drain 4 % + 4 % for vacancy (due to death of tea plants) = 8 %

After deducting 8 % , total requirement of MOP = 6921 kg.

So in this case , the split and the quantum of MOP for 20.5 hectares should be as follows :

1st Split – March-April – 6921 kg of MOP x 40 % = 2768 kg

2nd Split – June 1st week – 6921 kg of MOP x 30 % = 2076 kg

3rd Split- Mid September – 6921 kg of MOP x 30 % = 2076 kg

REQUIREMENT OF PHOSPHATE

Generally it is said that Phosphorus should be applied in the form of Rock Phosphate or Single super phosphate in the pruned and Deep Skiff years. However, if fund permits, it is better to apply phosphate every year. The dose of P_2O_5 is 50 kg/per hectare. In case of Rock Phosphate which contains 20% P_2O_5, the dose is 250 kg Rock Phosphate per hectare. In case of Single Super Phosphate , the dose of Single Super Phosphate is 312.5 kg per hectare.

WHICH KIND OF PHOSPHATE SHOULD BE APPLIED?

Single super phosphate(SSP) is a water soluble phosphate and Rock phosphate(RP) is an acid soluble phosphate. Hence, if the pH of the soil is less than 5.5 then Rock phosphate should be used and if it is more than 5.5 then SSP should be used. Rock phosphate is a slow releasing phosphate and found to be effective in the mature tea bushes.

Di Ammonium phosphate is given as foliar application in tea gardens @ 22 kg per hectare per year at 1 % concentration each time.

BASAL OR SPLIT : PLACEMENT OR BROADCAST IN CASE OF PHOSPHATE APPLICATION

Phosphate is immobile in soil and leaching loss is almost zero. In few experiments, it has been noticed that placement of phosphatic fertilizers gives better result than broadcasting. Placement method is digging a small pit and covering with soil after putting the fertilizer.

In case of Light pruned and Medium pruned sections, it is better to apply phosphatic fertilizer in basal dose (once in a year) by placement method. It is better to dig a small pit in between two pruned bushes , put 30 grams of Rock phosphate and cover with soil.

In the skiffed sections , it is difficult to dig pits in between two plants because the top surface is covered up. Hence, in the skiffed sections, phosphatic fertilizers should be applied by broadcast method along with Nitrogen and Potash.

So in this case , the split and the quantum of Rock Phosphate for 20.5 hectares should be as follows :

Total requirement of Rock Phosphate is 250 kg per hectare (RP contains 20 % P_2O_5) X 20.5 hectares = 5125 kg. After deduction for vacancy + road/drains – 4% + 4 % (due to death of tea plants) = 8 % , it comes to 4715 kg for 20.5 hectares.

1st Split – March-April – 4715kg of RP x 40 % = 1886 kg

2nd Split – June 1st week – 4715 kg of RP x 30 % = 1415 kg

3rd Split- Mid September – 4715 kg of RP x 30 % = 1415 kg

SULPHUR

Earlier the fertilizer programme used to made of NPK, but now the concept has changed, now it is NPKS. Now a days , along with NPK, Sulphur has also been made an essential component of fertilizer programme because Sulphur plays a great role in increasing flavor of tea . It contains three amino acids (cystine, cystiene and methionine) which serve as the precursor of VFC synthesis.

The dose of Sulphur is 40 kg Sulphur per hectare per year to be given in two splits, pre-monsoon and post- monsoon alongwith NPK. . Sulphur should not be applied during rainy season because it will evolve Hydrogen Sulphide which can be toxic to tea plants. When elemental Sulphur (which contains 80 % Sulphur) is applied , in each split 45 kg of Elemental Sulphur should be applied per hectare.

Total requirement of Elemental Sulphur for 20.5 hectares is 1025 kg. After deduction for vacancy + road/drains – 4% + 4 % (due to death of tea plants) = 8%, it comes to 943 kg for 20.5 hectares

So in this case, the split and the quantum of Elemental Sulphur for 20.5 hectares should be as follows:

1st Split – March-April – 943 kg of Elemental Sulphur x 50 % = 472 kg

2nd Split – June 1st week – No application

3rd Split- Mid September – 943 kg of Elemental Sulphur x 50 % = 472 kg

FINAL CALCULATION (FOR 20.5 HECTARES) (IN CASE OF THREE SPLITS)

Urea	1ST SPLIT (40%) 3606 kg	2ND SPLIT (30%) 2704 kg	3RD SPLIT (30%) 2704 kg
MOP Rock Phosphate	2768 kg 1886 kg	2076 kg 1415 kg	2076 kg 1415 kg
Elemental Sulphur	1ST SPLIT (50%) 472 kg	2ND SPLIT (0%) 0	3RD SPLIT (50%) 472 kg

Now, it is required to calculate the fertilizer per plant. First, the number of plants per hectare to be calculated using the following formula:

Number of plants per hectare = (N x 10,000 square metres)/ Y metres (X metres + Z metres)

(Where N= Number of hedges

X = Distance between hedge to hedge

Y = Distance between plant to plant

Z = Distance between row to row

In case single hedge N= 1 and in case of double hedge N = 2)

In this case:

No. of plants per hectare = (1 x 10,000 m^2) / 60 cm (0 + 105 cm)

(here X=0 because there is only 1 hedge)

= (1 x 10,000 m^2)/ 0.6 metre (0 + 1.05 metres)

= 15873 plants /hectare

Total number of plants after deduction for vacancy and roads/drains etc (total 8 %) per hectare = 14603. So, for 20.5 hectares, it comes to 299361 plants.

FERTILIZER PER PLANT

1ST SPLIT : 8732000 g / 299361 plants = 29 g / plant

2ND SPLIT: 6195000 g / 299361 plants = 21 g / plant

3RD SPLIT: 6667000 g / 299361 plants = 23 g / plant

TO INCREASE SOIL pH:

Dolomite Doses (80-100 mesh size)

pH Range	Dose of Dolomite per hectare (in tones)		Time of application	Next Manuring
	Coarse textured soil	Fine textured soil		
Below 4.5	2	3	January	At least after 8 weeks
4.5 to 4.65	1	1.5	January	

In the shade tree planting pit, addition of 1000 g of Dolomite can be very useful.

APPLICATION METHOD OF DOLOMITE

Dolomite should be applied after forking of the interspaces. That is the reason that dolomite can not be applied during skiffed years. It should be applied after light or medium pruning when the interspaces are adequately exposed. A light irrigation after dolomite application really helps.

TO REDUCE pH

In many areas of Dooars, there is huge increase in pH because of Dolomite influx from the hills. To reduce pH, Sulphur/Iron Pyrites or Aluminium Sulphate is used.

SULPHUR/IRON PYRITES OR ALUMINIUM SULPHATE ?

Aluminium sulphate is easily available to the tea plants but Pyrites need microbial action for coming to the available form, hence, where the organic matter content in the soil is below 1 % , i.e. where the microbial population is less, it is suggested to apply Aluminium sulphate and not pyrites.

SULPHUR IN PLANTING PITS

Where the pH is too high, it is recommended to apply sulphur in the planting pit at the rates mentioned below:

Soil ph values	Sulphur per hole (g)
5.9-6.4	115
6.5-6.9	225
7.0-7.4	340
7.5 and more	450

(For 115 g of Sulphur , 450 g of Aluminium Sulphate is required)

In hut site areas where tea is not thriving well, Aluminium Sulphate @ 450 g per square metre placed on the ground every three months a year can be useful.

SOIL pH AND AVALABILITY OF NUTRIENT

Nitrogen: High pH – Availability increases

Low pH – Availability decreases

Phosphorus: In very acidic soil Phosphate gets bounded with aluminium, iron and calcium and forms aluminium phosphate, iron phosphate and calcium phosphate, hence, it becomes unavailable to the tea plants. Use of Phosphorus Solubilizing Microorganisms (PSM) or Phosphorus Solubilising Bacteria (PSB) like *Bacillus megatherium* and Vesicular Arbuscular Mycorrhizae (VAM) like *Rhizophagus thea* and *Protomyces thea* help in conversion to the available form.

Potassium, Calcium, Magnesium, Sulphur : With the decrease of pH, the availability of these nutrients decreases.

Aluminium, Iron, Manganese : At low pH also these nutrients are easily available to tea plants.

Copper, Zinc, Boron : High pH – Availability decreases

Low pH – Availability increases.

Molybdenum: In the acid soil Mo is precipitated with iron and aluminium. At higher pH , its availability increases.

TEA BOARD'S RECOMMENDATION ON THE RANGE OF pH & NUTRIENTS

pH – 4.5 to 5.5

Nitrogen – 0.1 to 0.2 %

Organic Carbon – 1 to 2 %

Phosphorus (P_2O_5) – 10-20 ppm

Potash (K_2O) – 60-80 ppm

Sulphur – above 30 ppm

FOLIAR APPLICATION OF NUTRIENTS

Foliar application means applying nutrients directly on the foliage of tea plants by sprayers after dissolving in water.

1. Application of Urea is a general practice in tea. Almost every month Urea is applied @ 2 kg per drum (1% concentration)
2. Di ammonium Phosphate (DAP) is applied , pre monsoon and post-monsoon at 1 % concentration totaling to about 22 kg per hectare per year.

3. Zinc Sulphate is applied about 12.5 kg per hectare per year in many splits @ 1 % concentration. The efficacy of Zinc increasing with addition of Manganese @ 300, Boron @ 100 g and Alpha naptha Acetic Acid (NAA) @ 45 ml per drum.
4. Spray of n-Triacontanol, a long chain fatty acid alcohol has proved to be very useful in tea.
5. In many tea gardens gibberellic acid is applied once in a year , preferably in the month of May during dormancy. Gibberellic acid is applied @ 4 g per drum (200 litres of water). Since gibberellic acid is insoluble in water, 4 g of it is dissolved in 8 ml of Acetone and then transferred to a 200 litre-drum.
6. Spray of a combination of Humic Acid and Fulvic acid formulation@ 1 % concentration is found to be very effective during lull (dormant) period and during drought.
7. In many tea gardens, amino acids are sprayed @ 1 % concentration for vigorous growth of tea flush.
8. Need based application of Magnesium Sulphate, 8 kg in a year @ 1% concentration is given.
9. During water stress period, spray of NK foliar (1 % Urea and 1% Muriate of Potash) helps the tea plants to fight drought. Potassium regulates the function of stomata and induces stomatal resistance during dry period.

ROLE OF n-TRIACONTANOL IN TEA

- Promotes photosynthesis.
- Depresses photo respiration.
- Keeps the cell turgid and induces succulence. Enhances water use efficiency.
- Activates efficient use of applied nutrients.
- Triacontanol is cost-effective and may increase yield upto 40 %.
- Triacontanol induces primary, secondary and tertiary growth.
- Tria increases polyphenol content.

Exercise No. 3

Vegetative Propagation

NURSERY BEDS

SOIL

pH should be 4.5to 4.8 (soil pH beyond 5 tends to promote excessive callusing at the expense of rooting). At higher level of pH, the roots of young plants die back (bitten off disease)

SLEEVE SIZE

Lay flat	-	6″ – 7″ (15 – 17.5 cm)
Length	-	9″ – 10″ (22.5 – 25 cm)
Gauge	-	150

BED SIZE

1.2 m X 30 m (2500 sleeves approx)

SOIL IN THE SLEEVES

The lower portion of the sleeve to be filled with jungle top soil, rich in organic matter or soil : FYM :: 4:1 combination.

The top portion to be filled with rooting media comprising of sub soil.

INPUTS

Following ingredients are added to the soil sleeve before planting of cuttings :

Trichoderma formulation– 1 -2 g + VAM (Vescicular Arbuscular Myccorhizae) – 5 g per sleeve

SINGLE SUPER PHOSPHATE

Add 1 kg of SSP per cubic metre of soil.

DRAINS

Between beds	:	Depth	-	45 to 60 cm
		Width	-	30 to 45 cm
Periphery	:	Depth	-	90 cm
		Width	-	60 cm

TIME OF PLANTING CUTTINGS

SPRING PLANTING : April / May

AUTUMN PLANTING : September to November

SOIL TESTING & SOIL TREATMENT

Soil analysis for pH and eelworm is necessary. pH should be between 4.5 to 5.2. If the pH is high apply 50 ml of 2% Aluminium sulphate solution per sleeve. Equal quantity of water to be applied twice to leach down excess quantity of Aluminium Sulphate and obtain uniform pH. To make 2% solution of Aluminium Sulphate, dissolve 2 kg of chemical in 10 litres water and boil, then make the solution upto 100 litres.

Eelworm should not be more than 6 per 10 g of soil. If eelworm is more adopt the following method :

Calculate the quantity of soil required to fill 1000 polythene bags. Each time take this quantity of soil and mix nicely 1 kg of Carbofuran granules before filling the bags.

SHADE

It is better to build overhead shade under which people can move freely. The height to be adjusted accordingly. The best thing would be to pt on Concrete pillars and Overhead structure and cover with netlon materials. The side walls should also be a netlon.

ARCHED POLYTHENE TENT (OPTIONAL)

During Winter construct an arched transparent polythene tent, 1 – 8 m wide, over the sleeves. The bamboo frame should be 75 cm high. Polythene sheet of 400 gouge generally used. These sheets do not become opaque for 2 – 3 seasons. The sheet should not bear any leak.

The tent maintains high humidity inside and regulates temperature. During drought water the tent regularly from outside. The tent may be opened once in a fortnight (or ten days) and the bed may be watered inside.

UPKEEP OF MOTHER BUSHES

Pinch off two leaves and a bud of the primaries to induce axillary buds, two to three weeks prior to harvesting the crop. This should be followed by foliar spray of 1% Urea solution and 2% aqueous solution of Zinc Sulphate (2 applications).

TIME OF PLANTING CUTTINGS

SPRING – April / May

AUTUMN : September/October/November

TIME OF PRUNING/ SKIFFING OF MOTHER BUSHES

LIGHT PRUNE	TIME OF TAKING SCIONS
Mid October to 1st Week of November	End April to Early June
Mid November to 1st week of December	Mid May to End June
1st week of January to Early February	June – July

DEEP SKIFF

Early June	September / October
Early July	October/November

PREPARATION OF CUTTINGS

The shoots from which the cuttings will be taken, should have the cut ends dipped in the water in a bucket and covered with fresh green plant materials. Green semi hard wood, one leaf cuttings consisting of One full leaf and an internode should be taken with a sharp knife. After each cut the cut end should be dipped in water. Immerse the separated cuttings in a bucket of water.

PLANTING OF THE CUTTINGS

The sleeves should be covered with black polythene sheet immediately after filling. In the evening, prior to putting out of the cutting, take out the cover apply 17 litres of water per square metre and again cover with polythene sheet.

Take the cuttings in the next morning. Punch small holes with a small stick, insert the cuttings gently. But before that dip the cutting (cut end) in .1% Zinc Sulphate solution. Do not dip the cuttings in Blitox solution. The node and the petiole should not be buried in the soil. The cutting should be planted in such a way that the leaves remain erect. The soil around the collar should be pressed hard to avoid air pockets. A light watering after planting of cuttings would be useful.

FERTILIZER APPLICATION

When the plant puts forth 3 to 4 leaves apply YTD

(N:P:K: : 10:5:10) mixed with soil / Sand (1:9) @ 5 g per sleeve at 3 weeks interval.

FOLIAR SPRAY

On the plants having more than 4 leaves, spray :

FIRST ROUND

DAP - 2000 g + Any Nursery Special formulation – 500 ml per 200 litres of water

SECOND ROUND

Urea - 2000 g + MOP-2000 g + Humic acid formulation– 200 ml per 200 litres of water

REASONS FOR POOR DEVELOPMENT OF PLANTS IN THE NURSERY

SYMPTOMS	REASONS
1. High rate of mortality : brown stem	Dryness/ inadequate shade
2. High rate of mortality : soft bark	Wet bed / water logging**
3. Cutting live but no root	Weak cuttings/ deep shade/water logging high pH
4. Over callusing and weak root	High organic matter / high pH / water logging.
5. Poor growth : Yellow leaves	Nematode infestation
6. Poor growth : pale green leaves	Low nutrient / water logging
7. Poor growth : thick , hard leaves	Dry soil
8. Small sickle shaped leaves/ rosetting	Zinc deficiency of leaves

** Reasons for water logging : a) Poor quality of polythene sleeves

b) Shallow drainage

c) Too much of watering

TIME FOR DEVELOPMENT OF CALLUS & ROOTS

Generally the cuttings take 4-6 weeks for callusing and 10-12 weeks for rooting under favourable conditions.

SHADE TREE NURSERY

BED PREPARATION

It should be done during November-December. Deep ploughing followed by harrowing will be required. Loosen the soil up to a depth of 60 – 70 cm. On this soil, heap up dry jungle litters and burn.

SSP

After preparation of soil apply Single Super Phosphate @ 125 g per square metre of soil

(If directly planted on the beds)

OTHER INPUTS

Well rotten cattle manure @ 25 MT and Dolomite (100 mesh) @ 2 MT per hectare. (If directly planted on the beds)

DRAINS

Between beds : Depth – 60 cm ; Width – 45 cm

Periphery : Depth – 120 – 150 cm; Width – 60 cm

SIZE OF BEDS : Width 150 ; Length – As per availability of area.

POLYTHENE SLEEVES : Size – 30 cm lay flat ; Length – 60 cm ; Thickness – 300 gauge

SOIL MIXTURE FOR SLEEVES

Friable Soil – 4 parts ; Cattle manure – 1 part; SSP – 1kg per Cubic meter of soil; Dolomite – 0.5 kg per Cubic meter of soil

VAM – 5 g + *Trichoderma viride*– 1 -2 g per sleeve.

SOWING

Soak the seeds in water overnight followed by sun drying before sowing. Seeds with very hard seed coat can be treated with either hot water (80° C) or Concentrated H_2SO_4 solution. When treated with hot water, immerse seeds in water and heat upto 80° C and allow to cool overnight. If H_2SO_4 is used, soak for only 1-2 minutes followed by thorough washing in running water.

TIME OF SOWING SEEDS

February to April

METHOD OF SOWING

DIRECTLY ON THE BED

Depth – 0.5 to 1.0 cm

Spacing – 20 cm

Interline spacing – 30 cm

Cover with thin layer of soil and water lightly

SLEEVES

Drill 2-3 seeds in each sleeve at a depth of 0.5 to 1.0 cm. Cover with thin layer of soil and water.

PEST CONTROL

Soak the beds with insecticides after sowing. Pests to be controlled as per requirement.

Exercise No. 4

Young Tea Management

NEW PLANTING – Planting on the virgin land, where there was no tea plants earlier.

REPLANTING – When a tea plantation becomes economically unviable due to age or other reasons, replanting is done by complete removal of old plants and planting fresh lot of tea plants. Generally after 60 years of age, a tea plant does not remain economically viable, hence replanting is done.

REPLACEMENT – When a registered area of a tea plantation becomes unusable due to land slide or dolomite influx, the similar area of land in a virgin area is planted with young plants, this is known as replacement planting.

FOR REMOVAL OF EXISTING OLD BUSHES

There are many methods employed for uprooting / removing old tea bushes.

Manual – The bushes are uprooted manually. In the hilly region where slope percentage is high it is difficult to uproot by using any machine, hence uprooting is done by hand. A worker can generally uproot 30-40 bushes per day.

No-tillage method – The collar is chopped off and on the chopped collar, 500 ml of 2,4-D solution is poured (1 kg 2,4-dicholoro phenoxy acetic acid in 75 litres of water). At some places, Tractor winch / Bulldozer / Crane JCB – In the plains, the uprooting is done by any of these machines and these are found to be very effective.

As per Tea Board's regulation, in case of replanting, rehabilitation of the soil should be done for a minimum of 18 months after removal of old bushes.

REHABILITATION

Plants (cover crops) suitable for rehabilitation :

Tripsacum laxum (Guatemala grass) – The lopping of the plants adds to the humus of the soil.

Mimosa invisa (Thornless Mimosa)- Being leguminous in nature it has the capacity to fix atmospheric Nitrogen.

Pennisetum purpureum (Pusa Giant hybrid Napier grass)- The lopping of the plants adds to the humus of the soil.

Tripsacum laxum Produces 100 tonnes of green matter per hectare. 10000 to 12000 cutting are required for one hectare. In highly acidic soil 2 tonnes of Dolomite (a mixture of Calcium carbonate and Magnesium carbonate) per hectare is broadcast 8-10 weeks before planting.

At the time of planting N: P: K : 10:5:10 @ 600 kg per hectare and 300 kg per hectare after lopping are applied.

Lopping – After the plant reaches 25-30 cm height, lopping is done at 8-10 weeks interval. Last lopping should be in October.

For one hectare plantation. 0.2 hectare nursery is required.

Pennisetum purpureum Produces 200 tonnes of green matter per hectare. 10000 to 12000 cutting are required for one hectare. In highly acidic soil 2 tonnes of Dolomite (a mixture of Calcium carbonate and Magnesium carbonate) per hectare is broadcast 8-10 weeks before planting.

At the time of planting N: P: K : 10:5:10 @ 600 kg per hectare and 300 kg per hectare after lopping are applied.

First Lopping should be done 16-20 weeks after planting.

For one hectare plantation. 0.1 hectare nursery is required.

Mimosa invisa 3-4 kg seeds required per hectare. Rows should be 1.2 metres apart. Dolomite 2 tonnes per hectare should be broadcast 8-10 weeks before sowing. Single super phosphate @ 1 kg per 25 running metres to be applied in the rows.

LAND PREPARATION

All the existing bushes and shade trees should be completely removed from the area to be planted. The roots should also be uprooted in case of normal uprooting method. In case of no- tillage replanting, the roots get decomposed by chemicals.

Topographic Survey : Before starting of drain digging, it is necessary to go for topographical survey which is also known as contour survey. From the topographical survey map, the plan should be drawn for proper drainage.

Drains : All category of drains viz main drain, sub main drain or collector drain and field drain or lateral drain should be dug before planting. Lateral drain and the row of tea plants should always be against the slope. The soil of the excavated drains should be spread properly so that there is no accumulation of soil along the drain side. The shape of the drain should be trapezoidal. In case of deeper drains, do not dig the whole depth at a time otherwise there can be huge soil erosion and the drains may collapse. It should be done in at least two phases. The dimension and spacing of drains should be as follows :

SPACING OF LATERAL DRAIN

Loamy sand	-	18 metres
Sandy loam	-	12 metres
Loam	-	9 metres
Silt loam	-	6 metres

DIMENSION OF LATERAL DRAIN

Drain	Drain depth	Top width	Bottom width
Lateral	105 cm	45 cm	15-20 cm
Sub-Main	120 cm	75-100 cm	30-50 cm
Main	150 cm	100-150cm	50-100 cm

The lateral drains should always be against the slope. At the meeting point of two drains, there should be a little curve so that soil erosion can be reduced.

Before digging the drains, criss- cross ploughing, criss-cross harrowing, criss- cross sub soiling

(if required) and levelling should be done. All the stone particles should be removed from the land.

After drain construction, lay out for planting should be done.

STYLE OF PLANTING AND SPACING

Single hedge – Regular (rectangular) or staggered (triangular)

Double hedge - Regular (rectangular) or staggered (triangular)

Now a days, single hedge is preferred because in the double hedge, movement of workers, plucking and application of inputs become difficult.

Formula to calculate number of plants per hectare:

Number of plants per hectare = (N x 10,000 square metres) / Y metres (X metres + Z metres)

Where N= Number of hedges

X = Distance between hedge to hedge

Y = Distance between plant to plant

Z = Distance between row to row

In case single hedge N= 1 and in case of double hedge N = 2

The best spacing is 105 cm (between rows) x 60 cm (between plants) staggered single hedge.

Examples :

SINGLE HEDGE PLANTATION

Say the spacing is 105 cm (distance between row to row) X 60 cm (distance between plant to plant). So putting the values to the equation following result will emerge :

Number of plants per hectare = (N x 10,000 square metres) / Y metres (X metres + Z metres)

= 1 x 10,000 m^2 / 60 cm (0 + 105 cm)

(here X=0 because there is only 1 hedge)

= 1 x 10,000 m^2 / 0.6 metre (0 + 1.05 metres)

= 15873 plants /hectare

DOUBLE HEDGE PLANTATION

Say the spacing is 105 cm (distance between hedge to hedge) X 75 cm (distance between row to row) X 60 cm (distance between plant to plant). So putting the values to the equation following result will emerge :

Number of plants per hectare = (N x 10,000 square metres) / Y metres (X metres + Z metres)

= 2 x 10,000 m^2 / 60 cm (105 cm + 75 cm)

= 2 x 10,000 m^2 / 0.6 metre (1.05 metres + .75 metres)

= 2 x 10,000 m^2 / 0.6 metre (1.8 metres)

= 2 x 10,000 m^2 /1.08 metres

= 20,000 m^2 /1.08 metres

= 18518 plants/hectare

STAKING

With the help of plantation chains, the sticks (bamboo sticks) are fixed. The row of the tea plants should always be against the slope and along the lateral drains. During putting the sticks, marking should be made for the permanent shade trees also. In case of tea plants single stick and in case of shade trees double cross sticks should be put. After marking with the sticks, the pits should be dug manually or with the help of mechanical pit digger.

PIT SIZE

The depth should not be less than 45 cm and the diameter should not be less than 30 cm.

INPUTS IN THE PLANTING PITS

Well rotten cow dung / Farm yard manure – 2 kg + Single super phosphate – 30 g. The cow dung or the FYM should be treated with carbofuran (in case of conventional garden) @ 2 g per kg of cow dung or *Trichoderma* powder (in case of bio-organic garden) @ 5g per kg of cow dung.

SOIL CORRECTION IN CASE OF HIGH pH

Soil pH value	Sulphur per pit (g)
5.9 to 6.4	115
6.5 to 6.9	225
7.0 to 7.4	340
7.5 and above	450

pH above 6 is in no way suitable for tea plantation.

Aluminium Sulphate

Places where organic matter content is less, it is better to apply Aluminum Sulphate @ 4.5 tonnes per hectare.

SOIL CORRECTION IN CASE LOW pH

Soil pH value	Dolomite (tonnes per hectare)	
	Coarse textured soil	Fine textured soil
Below 4.5	2	3
4.5 to 4.65	1	1.5

SPECIFICATION OF THE PLANTS FOR PLANTING

Height – should be around 40-45 cm.

Foliage – 12-16 leaves per plant

Stem Thickness – 0.5 to 0.8 cm

PLANTING TIME

Planting should be done pre monsoon or post monsoon. In the area which is prone to water logging, planting should be done post monsoon and in the high lying droughty areas pre-monsoon planting can be useful. It is better to plant either early in the morning or evening. It is better not to plant when there is scorching sunlight.

PLANTING METHOD

While planting, it must be ensured that the axillary bud is not dipped inside the soil. After planting, proper ramming is required so that no air pocket is created. The accumulation of water in the air pockets can lead to death of the plants.

MULCHING

Though water hyacinth and Guatemala grass are the best mulch materials, tea pruning litters, tea waste, Citronella , thatch and paddy straw can also be used for mulching. In many countries black plastic mulch is also used. For decomposition of the grasses, 100 kg of Ammonium Sulphate can be evenly spread on the mulch. Mulch should be kept at least 10 cm away from the plants to avoid heat damage.

BENEFIT OF MULCHING

- Mulching helps in conservation of soil moisture.
- It suppresses weed growth.
- It induces feeder root proliferation.
- It induces growth of microflora.
- It adds to the organic content of the soil.

NUTRITIONAL VALUES OF MULCH MATERIALS

Guatemala grass	N %	P %	K %
Tea pruning litters	2.5	0.5	1.5
Tea Waste	3.6	0.6	2.3
Paddy straw	0.5	0.07	0.6
Thatch grass	0.5	0.1	0.05
Guatemala grass	1.6	0.4	1.3

YOUNG TEA MANURING

YTD MIXTURE DOSES

$N:P_2O_5:K_2O :: 2:1:2$

Year of Planting	N (kg/ha)	YTD (kg/ha)	Method of Application
0 Year	20-40	200-400	Ring in 2-3 splits
+ 1 Year	80-100	800-1000	Ring in 4 splits
+ 2 Year	100-120	1000-1200	Ring in 4 splits
+3 Year	120-140	1200-1400	Ring in 4 splits
+4 Year	140-150	1400-1500	Strip in 2 splits

Note: Recommended by TRA

BRINGING UP OF YOUNG TEA PLAINS OF NORTH EAST INDIA

NON DROUGHTY AREA

PLANT AGE	MONTH	OPERATION	TIPPING
O YEAR	OCT- DEC	PLANT TEA & ALLOW TO GROW	
+ 1 YEAR	EARLY FEBRUARY- MID APRIL	PRUNE/DECENTRE AT 20 cm	TIP AT 60-65 cm ABOVE GROUND
+ 1 YEAR	JUNE JULY	REMOVE STRONG CENTRAL BRANCHES AND CRISS CROSS BRANCHES	
+ 1 YEAR	END OCTOBER	STEP UP BY A LEAF	
+2 YEAR	WHOLE YEAR	UNPRUNED	PLUCK TO JANAM
+ 3 YEAR	END JAN – EARLY FEB	1ST FRAME FORMING PRUNING (FFP) AT 35-40 cm	TIP AT 60-65 cm ABOVE GROUND
+ 3 YEAR	END OCTOBER	STEP UP BY A LEAF	
+ 4 YEAR	WHOLE YEAR	UNPRUNED	PLUCK TO JANAM
+ 5 YEAR	END JAN – EARLY FEB	FINAL FRAME FORMING PRUNING (FFP) AT 40-45 cm REMOVE THICK CENTRAL BRANCHES	TIP AT 65-70 cm ABOVE GROUND

BRINGING UP OF YOUNG TEA PLAINS OF NORTH EAST INDIA

DROUGHTY AREA

PLANT AGE	MONTH	OPERATION	TIPPING
O YEAR	APRIL - JUNE	PLANT TEA & ALLOW TO GROW	
0 YEAR	JULY – AUGUST	PRUNE/DECENTRE AT 20 cm	TIP AT 60-65 cm ABOVE GROUND
+1 YEAR	WHOLE YEAR	UNPRUNED	PLUCK TO JANAM
+ 2YEAR	END JAN – EARLY FEB	1ST FRAME FORMING PRUNING (FFP) AT 35-40 cm	TIP AT 60-65 cm ABOVE GROUND
+ 2YEAR	END OCTOBER	STEP UP BY A LEAF	
+ 3YEAR	WHOLE YEAR	UNPRUNED	PLUCK TO JANAM
+ 4YEAR	END JAN – EARLY FEB	FINAL FRAME FORMING PRUNING (FFP) AT 40-45 cm REMOVE THICK CENTRAL BRANCHES	TIP AT 65-70 cm ABOVE GROUND

BRINGING UP OF YOUNG TEA HILLY AREAS OF NORTH EAST INDIA

LOW ELEVATION

PLANT AGE	MONTH	OPERATION	TIPPING
O YEAR	JUNE - JULY	PLANT TEA & ALLOW TO GROW	
+ 1 YEAR	JANUARY - FEBRUARY	PRUNE/DECENTRE AT 15-20 cm REMOVE STRONG CENTRAL BRANCHES AND CRISS CROSS BRANCHES	TIP AT 55-60 cm ABOVE GROUND
+2 YEAR	WHOLE YEAR	UNPRUNED	PLUCK TO JANAM
+ 3 YEAR	JANUARY - FEBRUARY	1ST FRAME FORMING PRUNING (FFP) AT 30-35 cm	TIP AT 55-60 cm ABOVE GROUND
+ 3 YEAR	JULY - AUGUST	STEP UP BY A LEAF	
+ 4 YEAR	WHOLE YEAR	UNPRUNED	PLUCK TO JANAM
+ 5 YEAR	JANUARY	FINAL FRAME FORMING PRUNING (FFP) AT 40-45 cm REMOVE THICK CENTRAL BRANCHES	TIP AT 60-65 cm ABOVE GROUND

BRINGING UP OF YOUNG TEA HILLY AREAS OF NORTH EAST INDIA

HIGH ELEVATION

PLANT AGE	MONTH	OPERATION	TIPPING
O YEAR	JUNE - JULY	PLANT TEA & ALLOW TO GROW	
+ 1 YEAR	JANUARY - FEBRUARY	PRUNE/DECENTRE AT 15-20 cm REMOVE STRONG CENTRAL BRANCHES AND CRISS CROSS BRANCHES	TIP AT 55-60 cm ABOVE GROUND
+ 1 YEAR	EARLY SEPTEMBER	STEP UP BY A LEAF	
+2 YEAR	WHOLE YEAR	UNPRUNED	PLUCK TO JANAM
+ 3 YEAR	JANUARY - FEBRUARY	1ST FRAME FORMING PRUNING (FFP) AT 30-35 cm	TIP AT 55-60 cm ABOVE GROUND
+ 3 YEAR	EARLY SEPTEMBER	STEP UP BY A LEAF	
+ 4 YEAR& + 5 YEAR	WHOLE YEAR	UNPRUNED	PLUCK TO JANAM
+6 YEAR	JANUARY- MARCH	FINAL FRAME FORMING PRUNING (FFP) AT 35-40 cm REMOVE THICK CENTRAL BRANCHES	TIP AT 55-60 cm ABOVE GROUND

POST FFP TREATMENTS

1. Immediately after FFP, spray any systemic fungicide @ 1 g per litre of water on the cut bushes.
2. Knife Cleaning out to be followed within a week or so.
3. The interspaces should be forked using fork hoe to loosen the soil so that Oxygen enters the soil pore spaces which in turn will promote Nitrification process.
4. Organic manuring: In between two bushes, a pit should be dug and following ingredients should be inserted and covered with soil (placement method):
5. Well dried cow dung / Farm Yard Manure- 1 kg + Rock Phosphate – 30g per pit . FYM / Cow dung should be fortified using beneficial bacteria viz *Azotobacter, Azospirillum, Bacillus megatherium,* VAM etc. Now a days Organic manure fortifier are readily available in the market.
6. Mulching should be done using water hyacinth or thatch grass or Guatemala grass or any other green matter. The mulching should be thick enough to keep the soil under cover for a considerable period. The mulch should be at least 10 cm away from the collar region of the tea plants.

Exercise No. 5

Pruning

Pruning is one of the most important operations in a tea garden. The objectives of young tea pruning and mature tea pruning are different. In case of young tea pruning is done to induce lateral growth for producing more plucking points which in turn helps in increasing productivity. Mature tea pruning is done with the following objectives :

- To stop reproductive growth and encourage vegetative growth.
- To remove knots, unproductive diseased woods and change gnarled appearance of the bush.
- To provide a height which suits the pluckers.

TIME OF PRUNING

Mature tea : Mid- November to mid-January

Young Tea : End January to early February

(Maximum starch reserve found in February)

HOW TO DECIDE PRUNING TIME ?

1. **Movement of photosynthates**: Direction of movement of photosynthates from the maintenance foliage is decided by the growing stage of shoots. When the shoots are growing, the photosynthates move basically in upward direction to the growing shoots. But when the buds grow dormant when there are no growing shoots the direction of the movement of photosynthates is diverted towards the roots.
2. **Starch reserve and pruning time**: Starch reserved decides pruning time for the young and mature tea considering the bidirectional flow commencing October. This may be the earliest time for resting the bushes for prior to pruning another bidirectional movement occurs in May, which may be utilized for decentering in young tea. It may be noted that presence of good starch reserve in the roots is an important factor to reduce dieback and to have quicker and better recovery of bushes.

STARCH TEST

Iodine crystals (3 g) and Potassium Iodide (6 g) in one litre of water or weak hospital iodine solution may be painted on the newly cut, pencil thick root. Dark complexion reflects adequate starch reserve. If the colour does not change, it will denote that starch reserve is not adequate.

YOUNG TEA PRUNING

This aspect has been discussed in details in the YOUNG TEA MANAGEMENT CHAPTER.

TERMINOLOGIES OF PRUNING

FFP	=	Frame forming pruning
CP	=	Collar pruning
RP	=	Rejuvenation pruning
HRP	=	Height Reduction pruning
MP	=	Medium pruning (Cut back pruning)
LP	=	Light pruning (Cut across pruning)
DS	=	Deep Skiff
MS	=	Medium Skiff
LS	=	Light Skiff
LOS	=	Levelling – off - Skiff
GL	=	Ground Level

MATURE TEA PRUNING

HEIGHT OF PRUNING

Collar Prune : 10 cm from Ground level (GL)

Heavy Prune / Rejuvenation Prune : 40 – 45 cm in plains and 15 – 37 cm in hills from GL

Medium Prune : 50-65 cm in plains and 35 – 50 cm in hills from GL.

Height Reduction Pruning : 70 – 75 cm in plains only from GL.

Light prune : 4 – 5 cm above last pruning mark.

Deep Skiff : 10 – 13 cm above last LP mark.

Medium skiff : Just below majority of crow's feet.

Light skiff : At current years tipping level.

Levelling – off – skiff : 5 cm above current year's tipping level.

PRUNING CYCLES

IN THE PLAINS TEA GARDEN

3-4 years : LP-UP-UP ; LP-DS-UP ; LP-UP-DS ; LP-MS-DS ; LP-UP-DS / MS –UP

IN THE HILLY AREAS

4 – 5 years : LP-UP-DS-LS-UP ; LP-UP-UP-DS / MS –UP

TIPPING HEIGHTS

First plucking after pruning is known as Tipping

CP	:	70-75 cm from the ground level
MP/HRP	:	25-30 cm from the pruning mark
RP	:	25-35cm from the pruning mark
LP	:	20-25cm (avg. 5 leaves) from the pruning mark
DS	:	7-10 cm (avg. 2 leaves) from the pruning mark
MS	:	4-5 cm (avg. 1 leaf) from the pruning mark
LS / LOS	:	Janam (scale leaf) level

PRUNING KNIVES

For LP : 15 cm blade

For DS /MS : 25 cm blade

For LS / LOS : 30 cm blade

Cleaning Out : 7.5 – 10 cm blade

The weight of the LP knife should be not less than 450 g.

KNIFE CLEANING OUT (KCO)

1. All the criss cross branches , thin whippy branches, one of the two branches at the close proximity, hanging branches to be removed.
2. Old prune marks affected by *Poria hypobrunnea* should be scraped upto the clean wood.
3. A 7.5 cm blade hook type pruning knife should be used for KCO

BENEFIT OF BREATHERS

After LP / MP/HRP/RP it is better to keep a breather in each pruned bush.

- Breathers Provide shade to reduce sun-scorch.
- It removes harmful metabolites from roots.
- It mobilizes growth substances towards the production of new shoots.
- It contributes photosynthates to the roots.
- It influences the growth of feeder roots.

FEW IMPORTANT POINTS

- Cut marks should be as small as possible and round in shape.
- Cut marks should always face the center of a bush.
- Height of all the cut marks should be equal from the ground level.
- Cross branches, hanging branches, very thin, whippy branches, branches situated at a very close proximity must be removed.
- Always use measuring sticks at the time of pruning.
- Pruning litters should be preserved *in-situ*.
- Apply Bitumen paint on the bigger cut marks to prevent the infection of *Poria hypobrunnea* .

POST PRUNING TREATMENTS

1. Immediately after pruning , spray any systemic fungicide @ 1 gm per litre of water on the cut bushes.
2. Knife Cleaning out to be followed within a week or so.
3. The bushes should be completely defoliated.
4. The big cut marks should be painted with bitumen paint with the help of a brush.. Bitumen paint should be heated a bit before application. It should be applied during bright sunshine.
5. Lime wash of the pruned bush frame helps to clean mosses and lichens. It also helps to kill the eggs of the thrips in the crevices of the tea bushes. The proportion should be as follows :

   Quick lime – 1 kg + Caustic soda or washing soda – 3 kg to be dissolved in 50 litres of water, strained and sprayed on the pruned bushes with a hand operated sprayer.
6. The interspaces should be forked using fork hoe to loosen the soil so that Oxygen enters the soil pore spaces which in turn will promote nitrification process.
7. Soil correction is done during Light , Medium or lower pruning. If the pH is high , Aluminium Sulphate or Sulphur pyrites should be applied and if the pH is low, Dolomite should be applied on the interspaces after forking. A light irrigation should be given after application.
8. Organic manuring: In between two bushes, a pit should be dug and following ingredients should be inserted and covered with soil (placement method):

   Well dried cowdung / Farm Yard Manure- 1 kg + Rock Phosphate – 30gms per pit . FYM / Cowdung should be fortified / enriched using beneficial bacteria viz *Azotobacter, Azospirillum, Bacillus megatherium,* VAM etc.

9. The drains should be deepened or re graded as per requirement.
10. Thinning out of the shade trees , if required should be done after Light pruning or pruning lower than that.
11. The shade tree trunks should be painted with lime upto a height of 2 metres. 10 kg of quick lime + 500 gms of copper oxychloride (or any other fungicide) + 60 ml of thiacloprid (or any other insecticide) should be dissolved in 200 litres of water and the shade tree trunks should be painted with a brush.

MECHANICAL PRUNING

Looking at the scarcity of workers and increasing cost of manual pruning many garden has resorted to mechanical pruning with the help pruning machines with circular discs.

MECHANICAL PRUNING STATISTICS GATHERED FROM A TEA GARDEN OF TERAI IN JANUARY 2022 (THIS MAY VARY FROM GARDEN TO GARDEN)

COSTING PER BUSH		
	Manual	Mechanical
Light Pruning	65 paise/bush	41 paise /bush
Deep Skiff	40 paise / bush	23 paise / bush

MECHANICAL PRUNING MANDAYS REQUIRED

LIGHT PRUNING – 10 mandays /hectare

DEEP SKIFF – 7 mandays /hectare

Exercise No. 6

Plucking

PERIODICITY OF GROWTH

Tea plants have a rhythmic growth pattern. Freely growing tea shoots pass through alternating states of growth and dormancy. The amount of growth made between two states of dormancy is known as a 'flush'. In N.E. India, upto four flushes were recorded on unpruned seed trees in a growing season and upto five flushes on pruned bushes, although the fifth flush was rare and found only in a few genotypes.

The apex bud is tightly wrapped in a foliage leaf. Unfolding of the ultimate or 'banjhi' (dormant) buds indicate the beginning of a flush. These 'banjhi' buds are much smaller than growing buds. A 'banjhi' bud is enclosed by two or occasionally three cataphylls or bud scales known as 'janam' (birth scale) in N.E. India. A 'banjhi' bud does not show any visible sign of growth and may remain in that condition for the rest of the season.

INTERFLUSH DORMANCY

Interflush dormancy occurs in between two flushes i.e. in between 1st flush and 2nd flush, 2nd flush and rain flush.

CAUSES OF INTERFLUSH DORMANCY

a. It is caused by slow rate of maturation and development of tissues in the growing shoot. That is not able to keep pace with the fast growth and therefore the shoots go dormant,

b. The shoot also go dormant because of low availability of moisture from the soil and various other factors,

c. When number of shoots are more in the bush but the production of carbohydrate is less, the shoots go dormant.

Winter Dormancy: Winter dormancy occurs in between autumn flush and first flush.

CAUSES OF WINTER DORMANCY

a. The winter dormancy is caused by the interaction of some environmental factors like short day length and low temperature, which appear to be main factors influencing leaf period.

b. Tea is cultivated between latitudes 45 north and 35 south, and at or near the equator. It produces almost the same yield every month. But farther from the equator, the winter harvest gradually declines and at latitudes beyond about 16 there is almost complete winter dormancy which lasts longer with increasing latitudes.

c. Suitable temperature for growing tea is ranges from 12-13^{0}C to 28-32^{0}C. In winter lower temperature comes below 12^{0}C,which affect the hormonal regulation & ultimately causes winter dormancy in tea.

d. For proper growth and development of tea, day length should not be less than 11 hours 15 minutes. During winter, the day length is short or it is less than 11 hours 15 minutes, hence growth and development of young tea shoots are hampered and shoots go dormant.

e. Both in winter and inter flush dormancy it has been seen that with the change in the growing pattern of the shoot there is also a change in the endogenous growth promoters/inhibitors ratio. A high ratio makes the shoots grow and a low ratio make the shoots go dormant. According the direction of movement of photosynthates from the maintenance leaf also decided. The growth promoters are auxin, gibberelin, cytokinin and growth inhibitors are abscisic acid and ethylene. The balance is caused by environmental and genetic factors.

f. Both water deficit and excess water cause moisture stress to the plant. This moisture stress causes morphological, physiological and anatomical changes to the plant which sets dormancy in tea plants.

Dormancy of tea shoots decide the pruning time in tea. It is a good physiological characteristics of tea.

PLUCKING

Plucking in tea refers to harvesting of tender apical shoots growing above the pre-determined plucking table. Besides giving crop, it encourages regeneration of new shoots, checks vertical growth of bushes and keeps them in vegetative phase. Plucking accounts for 30-33% of total cost of green leaf production.

OBJECTIVES OF PLUCKING

Young tea – To encourage lateral growth and get some quantity of leaves.

Mature tea –i) To get economic crop for manufacturing. ii) To keep the plants in proper plucking height iii) To reduce dormancy and encourage new shoots iii) To reduce the infestation marks of pests and get more & proper leaf lamina which in turn enhances rate of photosynthesis.

PLUCKING ROUNDS

Plucking round is the time interval in days between two consecutive rounds. Plucking rounds are ascertained in two ways:

Phyllocron Method: (Leaf period x 2) – 1

Leaf period: Time taken by a 1+ apical bud to become 2 + apical bud, 2+ apical buds to become 3 + apical bud.

Thermal Time Regulated Plucking:

Under N.E. agrological condition, thermal time requirement for buds to become pluckable shoots is 389 day degrees.

$t = A / (T - Tb)$

Where, a) A = Thermal time requirement for buds become pluckable shoots = 389 day degrees.

b) T = Prevailing temperature = (maximum temp. + minimum temp.)/2

c) Tb = Base temperature = 12.5^{0}C

d) t = Time requirement by buds to become pluckable shoots (days)

STANDARD OF PLUCKING

Black plucking – Removes all the leaves and bud above tipping level except unopened buds.

Standard plucking- In this method all the leaves except apical bud and small one and bud are plucked.

Coarse plucking – One plus apical bud, small two plus apical buds are left and remainder is plucked off.

SYSTEM OF PLUCKING

Janam plucking – Plucking is done leaving janam on the flush. This is the most common form of plucking in the tea industry.

Fish leaf plucking – Plucking is done leaving janam and fish leaf on the surface. Generally fish leaf plucking is not done in this part. Only during the starting of the season, somewhere it is practiced.

Single leaf or step up plucking – Plucked above a normal serrated leaf. In this system janam, fish leaf and the mother leaf is left on the plucking table. After a prolonged drought or after a considerable period of water logging or a huge attack of looper caterpillars and red spider mites when top surface is defoliated due to huge senescence of leaves, this plucking is done.

This plucking is also known as leaf addition. The photosynthetic ability of a tea leaf remains up to 18 months but after 6 months a sharp decline in photosynthesis is noticed. Hence, there is a need to have a new set of leaves with full photosynthetic potential.

TIPPING

In a pruned bush, primary shoots arise from the dormant Buds on the pruned sticks. These are decapitated or tipped at pre-determined height to maintain the depth of Maintenance foliage and make a plucking table.

TIPPING HEIGHTS

LP = 20-26 cm (average height of 5 leaves)

DS = 7-10 cm (average height of 2 leaves)

MS = 4-5 cm (average height of 1 leaf)

LS = at skiffed level (janam plucking)

LOS= at skiffed level (janam plucking)

Collar pruning = 70-75 cm from the ground level

Rejuvenation pruning = 25-35 cm from the ground level

Medium pruning = 25-30 cm from the ground level

FINE COUNT BY BALLOMETRIC METHOD

Fine Count or Ballometric Count is a method which helps us to find the percentage of fine shoot in a flush. Fine leaves are responsible for getting good quality made tea. The fine leaf comprises of two leaves and a bud, small one leaf and a bud and soft single banjhi. It is done by breaking 6mm stem from the bottom of the second leaf.

FLUSH COMPONENT ANALYSIS

Flush component analysis gives information about the status of shoots in a flush. From this it can be ascertained whether plucking round needs to be extended or reduced, whether spray of hormones and growth promoting nutrients to be sprayed or not, whether there is problem in pruning cycle or plucking habit of the pluckers or not etc.

Process

Tender tea shoots are plucked and collected.

Shoots with active buds and shoots with dormant buds are selected.

From there 1+b, 2+b, 3+b, 4+b, 1+d, 2+d, 3+d, 4+ d single leaves are separated. Weight of 1+b is measured. By applying same method number and weight of other shoots are determined.

Active shoots				
	Number	Percentage	Weight	Percentage
1+ bud				
2 + bud				
3 + bud				
4 + bud				
Soft single banjhi				

Dormant (Banjhi) shoots				
	Number	Percentage	Weight	Percentage
1+banjhi				
1+banjhi				
1+banjhi				
1+banjhi				
Hard single leaf				

When a leaf is folded, if the bud is longer than half the leaf, it is an active shoot. If the bud is shorter than half of the leaf it is a dormant (banjhi) shoot.

Looking at the flush component analysis, the status of the plucked leaves is determined. On the basis of the status, root cause analysis is done and corrective action is initiated.

a. If the percentage of dormant shoot is more then it will indicate that the plucking round is extended or there may be hormonal problem or it may be the effect of bad plucking or adverse weather condition.
b. If the percentage of 1 + bud is more then it will indicate that the plucking round is too short.
c. If the percentage of hard single leaf is more, it will indicate that there is problem in plucking method.
d. If the percentage of 3 + bud is more, it will be considered as a perfect plucking under Dooars and Terai conditions.

HOW TO IMPROVE PLUCKERS' PRODUCTIVITY?

HIGH YIELDING SECTIONS: It has been observed that in high yielding sections

productivity of the pluckers is more Hence, proper seed and clone stock should be selected for plantation in accordance with agro-ecological conditions.

LEAF SIZE : Pluckers productivity would be more in the sections having tea plants with more leaf area. For example TV(Tocklai Vegetative) 30 has bigger leaf area and it is a high yielding clone.

HEIGHT OF THE PLANTS : Height of the plants should match the height of the pluckers. Hence, Anthropology has got something to do with the height of the tea plants. Proper pruning programme and control on creep can help. Longer plucking rounds also induces unnecessary creep. Hence, fix plucking rounds as per phylocron system.

FREE PASSAGE FOR THE PLUCKERS : There should be no obstruction on the way of the pluckers. Weeds, hanging branches and congested bush frames reduce pluckers productivity. Hence, it should be endeavoured to keep the section weed free. Hanging branches, if any should be trimmed as and when necessary. Lane cutting may also be practiced. Single hedge plantation also helps.

CATEGORISATION OF PLUCKERS ON THE BASIS OF PLUCKING : There are three types of pluckers – fast, medium and slow. Fast pluckers pluck 1/3 rd more than medium and slow pluckers pluck 1/3 rd less per unit time. If fast and slow pluckers pluck together, the slow pluckers ten to leave shoots on the bush surface to catch up with the fast pluckers. Hence deployment category should be taken care of.

DISTANCE BETWEEN PLUCKING AREA TO WEIGHMENT SPOT : Arrangement should be made to take weighment in the field itself very near to the plucking area, so that walking time is reduced.

LESS WAITING TIME : More spring balances should be employed so that waiting time for the shake of weighment can be minimized.

LESS LEAF CARRIAGE : By planning frequent weighment (say thrice a day), the burden of the pluckers may be lessened.Where three times weighment is not possible till the factory is installed, each plucker should be provided with at least two carrying bags (net / hessian)

ROLE OF SHADE TREES : During droughty summer days, the role of shade trees in improving pluckers' productivity is immense. Right kind of shade trees should be planted. It is imperative to plant shade trees which posses leaves during March –April. *Albizzia odoratisima* and *Derris robusta* would be the best choice. Shades like *Albizzia chinensis, Acacia lenticularis, or Albizzia lebbek* whose leafless period fall during March –April should be avoided.

CRECHE , proper arrangement for drinking water during plucking time and light music have been found effective in improving pluckers' productivity

STICKS: Both in pruned and unpruned sections, pluckers should use measuring " T " sticks. For prune / D.S. / M.S. sections a Tee may be prepared with extended arms to cover the entire plucking surface.

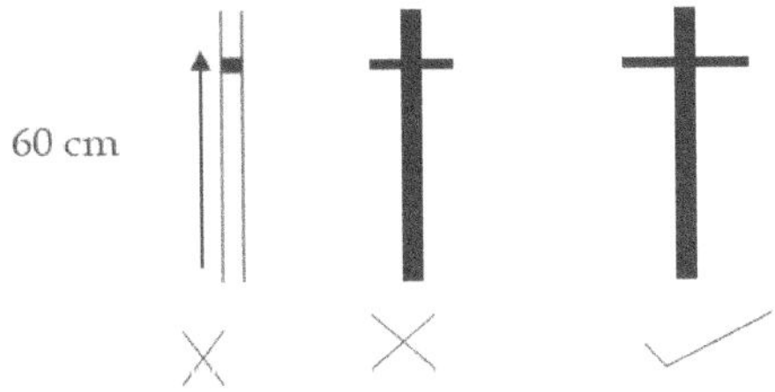

In skiffed sections a stick may be placed horizontally on the top surface so that the plucker can check their habit of dipping hands or pluck lowly placed side branches

MOTHER LEAF PLUCKING AND LEVEL PLUCKING: The plucker should be taught that the maintenance foliage which is the source of photosynthesis can retain peak photosynthetic ability maximum upto 6 months. Hence during the initial parts of the first flush and end of the season, mother leaf plucking should be practiced, while during peak growing season level plucking will help.

TRAINING OF THE SUPERVISORS: Training of the supervisors involved in plucking operation should be trained properly by AUDIO – VISUAL devices. They may also be sent to gardens with high profile of plucking .

PLUCKING SHEARS: To control plucking rounds and evenness of the surface during peak season (heavy flush), plucking shears may be introduced. Flat shears (30 cm length) should be used – not the two step shears.

PLUCKING TASK : The plucking task must be fixed very rationally, so that to fulfill the task the pluckers are not compelled to destroy the physiology of the tea plants

INCENTIVE: It has been observed that rational increase in extra – leaf price improves pluckers productivity.

In future the concept of THERMAL TIME REGULATED PLUCKING may be introduced.

MECHANICAL HARVESTING

In India about 67 % (about two-third) of the plantation workers are involved in plucking and about 33 % (about one-third) of the total expenditure in a tea plantation goes to plucking operation alone. In recent years there has been a sharp decline in the percentage of young workers in the tea plantation because the new generation off springs of the plantation workers are opting for career other than tea. In South India mechanical plucking is quite significant but in North East India, mechanical harvesting is yet to make in roads. To cope up with the crisis, it has become essential to introduce mechanical plucking in the tea industry. Mechanical plucking is also considered to be cost effective. Statistics on manual plucking and mechanical plucking is furnished below :

	1-man operated machine	2-men operated machine	Hand shears	Manual
Plucking Avg (kg) /worker	180	150	60	40
Total Plucking per day (kg)	450	750	60	40
Worker /hectare	6	8.33	15	20
Area Covered per day(hectare)	0.3-0.5	0.5-0.75	0.07	0.05
No. of machine /hectare	2-3	1.3-2	-	-
Plucking interval(days)	18-22	22-25	12-14	8-10

COST ECONOMICS/ HECTARE

	1-man operated machine	2-men operated machine	Hand shears	Manual
Labour cost(Rs)	512.5	1025	205	195
Fuel cost (Rs)	180	252	0	0
Maintenance cost (Rs)	350	600	2	0
Cost/kg	2.36	2.53	3.47	4.88

Note : Above statistics are generated by Tea Research Association, Tocklai Tea Research Institute.

REFERENCES

Barbora, B.C. Plucking and Leaf handling. In: The Planters' Handbook. Tea Research Association, Tocklai Experimental Station

Dutta, A.K., Saikia, D/N. and Sarma Jiten (2005) Tipping and Plucking of Tea.. In: Field Management in Tea. Tea Research Association, Tocklai Experimental Station, pp 90-113

Exercise No. 7

Drainage & Irrigation

Good drainage involves quick and safe disposal of run-off and soil water in excess of its field capacity from the crop root gone without causing problems of water logging and soil erosion.

BENEFITS OF DRAINAGE

Improves root and shoot growth and leaf harvest through:

a. Roots go deep into the soil.
b. Reduces soil erosion and harmful effects of drought
c. Soil keeps worm in winter and cool in summer, which is favorable for root growth.
d. Improves soil aeration, soil structure, microbial activity, decay of organic mater and soil fertility

PROBLEMS ASSOCIATED WITH WATERLOGGING

- Restricted out-fall.
- Run-off and seepage from higher catchment.
- Drain collapse.
- Sand boiling.
- Improper location and size of drains.
- Weed growth on drain bed and sides.
- Ponding due to uneven land surface.
- Subsoil impermeability.
- Unplanned roods, bunds, culverts etc.
- Wrong blacement of culverts.
- Overflow from streams, rivers etc.

Areas suffering from water logging give low yields, have poor and patchy growth, poor response of fertilizers, increased effect of drought, heavy weed growth and higher incidence of pests and diseases.

TYPES OF DRAINS

Subsidiary : These are laid down across the slope. (e.g. field, Lateral, Contour drains)

Primary : These are laid down along the slope. (e.g. Collector, Submain and Main drains)

Types of Drainage Systems

Random : Drains are dug out through natural depressions to drain out small, isolated low wet areas.

Herringbone : Where land slopes to middle from both sides, the collector drain is dug in the middle and the laterals join it from both sides.

Grid-iron :Laterals enter the callect6or drain only from one side.

Perimeter or Boundary Drain:_Prevents entry of water from outside areas.

Interceptor Drain : It is dug near the upper edge of wet area, where slope meets the plot land, to intercept seepage water.

Open Drains

Advantages:

Initial cost low

Easy to construct and maintain.

Can carry large quantities of water.

Disadvantages:

Occupy large areas which can otherwise be put under plantation.

Requires culverts and bridges.

More difficult and expensive to maintain.

Pipe Drains

Advantages:

- Useful in sandy areas.
- No loss of planting area.

- Culverts and bridges are not required
- Subsequent maintenance costs practically nil

Disadvantages:

- Initial cost very high.
- Requires technical expertise to lay down these drains.

CAUSES OF DRAIN COLLAPSE

- Sedimentation – seen more if bed gradient is inadequate and not uniform.
- Improper location and alignment of drains.
- Improper junctions.
- Inadequate shape and size of drain.
- Excessive weed growth in drains.
- Blockage of outlet drain.

DRAIN CONSTRUCTION

- Dig drains during dry weather (NOV-MAR/APR).
- Dig main drain first, followed by submain and lateral drains.
- Start digging from outlet and proceed upgrade.
- Dig in stages.
- Spread the dug-out soil evenly over the area, or keep it a little away on lower side of the drain
- Avoid digging when soil is wet.
- Carry out repair of collapsed drains in dry weather.

DRAIN MAINTENANCE

- Keep drain free of weeds, sediments etc.
- In sandy soils, grow deep rooting grass on drain sides.
- Avoid weeding on drain sides.
- Clean and re grade annually at least main/submain, drains.
- A good drain keeps flowing.

CULVERTS ON ROADS & PATHS

- Use hume pipes of max. 1.2m dia for paths, and 2.4m dia for roads. RCC Slabs can be also used.
- Join two pipes firmly end to end with a collar.
- Bury the hume pipes at least 15cm into the drain bed, and put a minimum 45cm soil cover on top.
- Align the pipes with the drain, not with the road.

SPILLWAY

This is required where vertical drop exceeds 100cm.

It may be made of bricks/stones in same way as slides in children's park.

DRAIN SPECIFICATIONS

Shape:_Should be trapezoidal or half-trapezoidal.

SLOPE (HORIZONTAL : VERTICAL)

Light soil (Loamy-sand)	0.5:1
Medium soil (Sandy Loam)	0.25:1
Heavy soil (silt loam, loam, clay loam)	0.25 : 1

SPACING OF FIELD DRAINS IN FLAT LANDS (SLOPE <3%) (metres)

Soil type	Assam	Dooars/Terai
Clay loam	9	6
Silty loam	12	9
Loam	15	12
Sandy loam	18	15
Loamy sand	30	25

DEPTH vs SPACING FOR MEDIUM TEXTURES SOILS

Depth	Spacing
105 cm	18 m
120 cm	28 m
135 cm	38 m
150 cm	48 m

Put the excavated soil on the lower side a little away from drain, or spread it evenly on ground.

SPACING ON SLOPY LANDS (SLOPE >3%)

In medium-textured soils, provide drains along contours as per following guideline:

Rainfall (cm)	**Vertical interval (m)**
>250	(% Slope +3) X 0.3048
251-400	(% Slope +2) X 0.3048
>400	(% Slope +1.75) X 0.3048

Soil type	**Adjustment (%)**
Coarse (Sand, sandy, loam)	+20to +4
Med. Coarse (Loam, Silt. Loam)	+4 to -4
Med. Fine (Silt-loam, silty-Cloy-loam)	-4 to -12
Fine (Silty-clay, Clay)	-12 to -20

Design Specifications

	Lateral Drain	Submain Drain	Main Drain	Perimeter Drain
Drain depth (cm)	105	120	150	150-180
Top width (cm)	45	75-100	100-150	150-250
Bottom width (cm)	15-20	30-50	50-100	150-250
Max.bed gradient (%)	0.25	0.15	0.10	0.25
Depth of flow (cm)	10	15	30	-
Max. flow velocity (cm/sec)	60-70	60-70	60-70	60-70

DRAIN JUNCTIONS

Drains should join together at an angle <90 and at bed level, with no out-fall.

DRAIN LOCATION

Should be as straight as possible or with gradual curve.

Use existing drains as much as possible.

Put also a perimeter drain, if possible.

◈ If measured by use of float, multiply by a factor of 0.8

SELECTION OF DRAINAGE SYSTEM

Flatland with slope up to 0.5%: Herring-bone system.

If outlet is restricted, 90 for pump drainage.

Land with mild slope of 0.5 – 1% : Grid-iron system.

Land with slope exceeding 1% : Provide interceptor

Drain across the main slope, and main drain down the main slope.

Undulating land with rolling topography: Grid-iron system of drains, with perimeter drain if required.

Land with basin-type topography: Requires pump drainage

Seepage from high lands : Deep interceptor drains across the main slope.

Water table fluctuates with rain fall : Improve surface and sub-surface drainage.

Soils gradually changing to more permeable strata with depth : Provide deep open/pipe drains.

Sandy loam and heavier soils : Provide suitably spaced open drains.

Sandy soils : Provide pipe drains

FEW DEFINITIONS ON DRAINAGE

Catchment : Area bound by ridges and discharging excess water at one point.

Out fall or outlet : The point to which w3ater is drained out.

Contour : Imaginary line drawn by joining points of same elevation with reference to certain datum line

Vertical drains : Wells dug through restrictive layer to drain water downward as in case of perched water table.

Relief Well : A shallow well carrying water under hydro-st5atic pressure from a subsurface layer into a drain.

Sump : Reservoir created to facilitate operation of pump.

Sand Boiling : Rising of bed surface due to pressure exerted by seepage water on light textured soil.

MAKING OF DRAINAGE SCHEMES

Determine the size of area to be drained

Carry out topographic survey and prepare contour map.

Contour interval : 30 cm for flat land

100 cm for steep topography

Scale for ≤120 ha – 1:1000, 121 to 240 ha – 1:2000,

>240 ha – 1: 4000

Based on topography, select type of drainage system and mark location of drains.

Study soil properties and decide about shape, side slope and bed gradient of drains.

Study quantum and pattern of RF distribution

Outlet availability, its elevation and distance.

Size and location of bridges, culverts, roads etc.

Average and high flood levels of drainage outlets.

DRAIN STABILTY

Level of under-ground water table and its fluctuations.

If required, LS profile of existing main/sub main drains.

For each catchment, develop main drains following natural water course starting from out6fall. Avoid sharp bends.

Provide collector drains along the slope, one for each sub-catchment.

Mark lateral drains across the slope in flat lands, and along contours in slope in areas.

Provide bed gradient from the highest point to the lowest point of outfall.

Master Row Planting

Location of master row is done on following principle:

- If a pair of graded contour drains converge in their direction of flow, the master row is made parallel to the top or higher contour drain.
- If a pair of graded contour drains diverge in direction of their flow, the master row is made parallel to lower contour drain.

Calculation of size of Main/submain Drains

Volume of Water Discharged

Q = 0.0028 C;A or Q = DCX A, where

Q = vol. of water discharged (m3/Sec)

C = run – off Coefficient (0.3 – 0.5, higher for heavy soil)

I = Intensity of R.F. (mm/hr)

A = Catchment area (ha)

DC = Drainage coefficient or depth of water to be removed (mm/sec)

Bottom width of Drain

A = bd + zd2, where

A = Area of water flow (sq.m.)

B = Bottom width (m)

D = Design depth of flow in drain (m)

Z = Drain side slope (horizontal/vertical)

Area of Water Flow

A = Q/V, where

A = Cross section of drain water flow (sq.m.)

Q = Volume of water discharged (cu.m./sec)

V = Velocity of water flow (m/sec)

FEW TIPS ON IRRIGATION

IRRIGATION SCHEDULE

FOR UNPRUNED AREAS

	Irrigation	Frequency
Nov	2″	once a month
Dec	2″	once a month
Jan	1″	once a month
Feb	1″	once a month
Mar	2″	once a month
Apr	4″	2″ fortnightly
May	6″	2″ every 10 days

FOR LP & DS AREAS

Jan	1″	once a month
Feb	2″	once a month
Mar	2″	once a month
Apr	2″	once a month
May	4″	2″ fortnightly

NOTE : 25 mm or 1″ rain equals 250 tons / hectare ; say 10000 l pump running for 25 hours

(Courtesy Tea Research Association)

FEW EQUATIONS ON SPRINKLER IRRIGATION SYSTEM

* Required discharge from an individual sprinkler (litres/second) = $\frac{SI \times Sm \times 1}{3600}$

SI = Spacing of sprinklers on the lateral line (mitres)

Sm = spacing of laterals on a main line (metres)

I = rate of water application (mm/hour)

- Total pump discharge = N X Q

N = No. of sprinklers run at a time.

Q = discharge from each sprinkler (litres/second)

- Discharge capacity of sprinkler system (litres/ second) = $\frac{2.78 \times A \times d}{F \times H \times E}$

A = Area irrigated

d = gross depth of each irrigation (mm)

F = No. of days allowed for completion of one irrigation

H = No. of actual operating hours per day

E = Irrigation efficiency (say 80%)

AREA COVERAGE PER SPRINKLER PER SETTING

24 m by 24 m	=	80′ by 80′	=	6400 sq.ft	=	0.147 acre
24 m by 30 m	=	80′ by 100′	=	8000 sq.ft	=	0.184 acre
30 m by 30 m						

TENSIOMETER

To know about when to start irrigation or when to stop irrigation, installation of tensiometers would be essential.

TENSIOMETER INSTALLATION

A pipe which is 2-3 mm less in diameter than the tensiometer tube diameter is used .

The metal pipe is hammered into the soil to a depth of about 10 cm less than the size of tensiometer and the pipe is pulled out with soil leaving a vertical hole behind.

A little water is poured into the hole to make the bottom soft to accept the tensiometer

After the water sinks into the soil, gently the tensiometer is pushed into the core vertically so that the porous cup firmly gets into the soft soil below the hole maintaining a close contact with the soil.

The soil around the tensiometer body is then tamped at the surface to seal around the body tube to prevent surface water from entering around the tensiometer body.

After installation , several hours required before the tensiometer reads the correct soil suction value. Protect the tensiometer from any damage by people and animals by constructing a box on the tensiometers.

TENSIOMETER READINGS

0 - 0.10 : it shows soil saturation.

– 0.2 : soil moisture in field capacity. The spriklers may be turned off at this point

– 0.5 : Good availability of soil moisture to plants. Irrigation not required

0.5 – 0.6 : It is the usual range for starting irrigation.

Higher than 0.6 : Moisture stress range

Exercise No. 8

Shade Trees

Shade trees have become the integral part of the tea gardens in the plains because of the following benefits:

1. It cut off solar insolation which is detrimental to tea plants.
2. It does not create high amount of darkness, it allows 50% of the sunlight to pass in.
3. Being leguminous plants (mostly from the family *Fabaceae),* their roots can harbour nitrogen fixing bacteria. This adds nitrogen in the soil.
4. Since leguminous shade trees are deciduous, it sheds leaves once in a year, fallen leaves adds humus and helps in nutrition.
5. It creates a micro climate in the phyllospheric region of the tea plants, conserve moisture and helps the plants to continue normal photosynthesis.
6. Regulated shade reduce the infestation of pests which are positively thermo and phototropic viz. mites, thrips etc.
7. Shade trees serves as wind/storm barrier.
8. The roots of shade trees are strong and highly proliferated, thereby it can hold the soil and reduce soil erosion.
9. There are many shade trees e.g. *Albizzia adoratissima,* wood of which is used for making furniture.
10. Twigs of shade trees are used as firewood.
11. Shade trees provide shade which increases the productivity and working energy of the pluckers and other workers during hot summer days.

CHARACTERISTIC FEATURES OF FEW MAJOR SHADE TREES

Albizzia chinensis (Osbeck) Merr syn stipulata Boiv

Family: Fabaceae

Common Name – Sau Sirish

Category : Semi-permanent shade tree

A medium sized tree. Maximum height about 15 metres. Bark colour is light brown.

Leaf : Compound leaf, bipinnate, paripinnate. Length of leaf rachis is about 30 cm with a gland at the base of petiole. Large stipules, about 2 cm long is the characteristic feature of this plant. Pinnae 10-20 pairs, 5-18 cm long. Leaflets are sharp pointed at the apex. Leaflets are sessile, 30-50 pairs , 0.7 to 1.5 cm long and 0.2 to 0.4 cm broad.

Fruit : Length of the pod is 12-15 cm and width is 1.5 to 2 cm. . The colour of the pod is straw colour when ripe. In each pod there are 4 to 10 seeds present.

Leafless Period : End January to end March. Hence this shade tree is not very suitable in a droughty area,

Density of shade : light to medium

Albizzia odoratissima Benth.

Family : Fabaceae

Common Name : Kala Sirish

Category : Permanent shade tree

A large deciduous tree with spreading crown. The colour of the bark is grey.

Leaf : Pinnately compound , bipinnate, paripinnate. Length 10 to 25 cm. Pinnae 4 to 8 pairs. Each pinnae 8 to 15 cm long. Leaflets 6 to 16 pairs ,dark green in colour, 1 to 2 cm long and 0.4 to 0.7 cm wide.

Fruit : Length of the pod is 15 to 30 cm and width 3 to 4 cm. The colour of the pod becomes dark brown when ripe. . In each pod there are 8 to 12 seeds present.

Leafless period : End December to February. This shade tree is very suitable in the droughty areas. In fact this should be the first choice as a shade tree.

Derris robusta Benth.

Family : Fabaceae

Common Name : Derris, Sirkhi

Category : Permanent shade tree.

A medium sized tree . The height goes upto 15 metres. The colour of the bark is whitish grey.

Leaf : Simple unipinnate, imparipinnate. Length 15 to 20 cm. Leaflets 10-25. Length of each leaflet is about 2 to 3.5 cm and width about 1 to 2 cm. Apex of each leaflet is mucronate.

Fruit : The length of the pod is 2.5 to 5 cm and 0.5 to 0.7 cm wide. The pods become straw coloured when ripe. Number of seeds in each pod is 1 to 4.

Leafless Period : Mid December to February. This shade tree is very suitable in a droughty area.

Sometimes the *Derris robusta* plants serve as the host of looper caterpillars but if proper pest control measure is taken, this species of shade tree can be very useful in a tea garden.

Albizzia lebbek **(L) Benth.**

Family : Fabaceae

Common Name : Lebbek, Lasrin

Category : Permanent shade tree.

A medium sized tree going upto the height of 15 metres. The colour of the bark is light grey to dark grey.

Leaf : Pinnately compound , bipinnate, paripinnate. Length 7 to 15 cm. Pinnae 3 to 6 pairs. Each pinnae 10 to 15 cm long. Leaflets pale green in colour, 3 to 8 pairs , 2 to 5 cm long and 1 to 2.5 cm wide.

Fruit : Length of the pod is 15 to 30 cm and width 3 to 4 cm. The pod becomes straw coloured when ripe. In each pod there are 6 to 10 seeds present.

Leafless period : January to April. This shade tree is not very suitable for the droughty area.

Acacia lenticularis **Buch-Ham.**

Family : Fabaceae

Common name : Kanta sirish.

Category : Permanent shade tree

The tree is about 15 to 20 metres high. The characteristic feature of this shade tree is it has hooked spines. The colour of the bark is grey.

Leaf : Pinnately compound, bipinnate, paripinnate. Length 12 to 20 cm. Pinnae 12 to 16 pairs. Each pinnae 10 to 20 cm long. Leaflets pale green in colour, 12 to 16 pairs, 2 to 2.5 cm long and 0.7 to 1 cm wide.

Fruit : Length of the pod is 15 to 20 cm and width 2.5 cm. The pod becomes dark brown coloured when ripe. In each pod there are 5 to 8 seeds present.

Leafless period : March to April. This shade tree is not very suitable for the droughty area. But this shade tree is very suitable in the area which has cattle trespass problem.

Adenanthera pavonina Linn.

Family : Fabaceae

Common Name : Rakto chandan

Category : Permanent shade tree.

The height of the tree 5 to 10 metres. The colour of the bark is brownish grey.

Leaf : Pinnately compound, bipinnate, paripinnate. Length 20 to 40 cm. Pinnae 3 to 6 pairs. Each pinnae 7 to 10 cm long. Leaflets 6 to 16 pairs , 2 to 3 cm long and 1 to 2 cm wide.

Fruit : Length of the pod is 15 to 20 cm and width 1.5 cm. The pod splits up when ripe. In each pod there are 8 to 16 seeds present. The colour of the seeds is bright scarlet red .

Leafless period : January to March

Indigofera teysmanni Miq.

Family : Fabaceae

Common Name : Indigofera

Category : Temporary shade tree

This is a small tree about 4 to 5 metres in height.

Leaf : Pinnately compound, unipinnate, imparipinnate. Length 20 to 30 cm. Leaflets lanceolate, apex and base tapering, 8 to 15 pairs, 5 to 8 cm long and 3 to 4 cm wide.

Fruit : Pod is cylindrical. Length of the pod is 2,5 cm and width 0.3 cm. The pod becomes blackish brown coloured when ripe

Leafless period : Mid January to Mid March.

Exercise No. 9

Bio Organic Farming / Biological Control of Pests

The term 'Bio-Organic Farming' is used for the plantations where no synthetic chemical pesticides and inorganic fertilizers are used. It is better to term as 'Bio-Organic Farming' rather than only ' Organic Farming' because most of the synthetic pesticides are organic in nature. Although very low in number, there are many tea gardens where the concept of bio-organic farming is strictly followed.

Synthetic pesticides and inorganic fertilizers induce environmental pollution and health hazards. Although through various seminars , various publications it has been endeavoured to generate awareness among people about menace of synthetic chemicals, the level of reduction in chemical use is still insignificant. After the publication of the book 'Silent Spring' by Rachel Carlson , there has been a growing consciousness among the planting community to reduce the use of conventional synthetic chemical pesticides and keep the residue level under control. At one point of time, it was almost impossible to replace conventional pesticides and fertilizers because alternatives were hardly available. But today, host of alternatives are available in the market. This revolutionary change in attitude of the manufacturing companies has ushered in a new era in easy conversion of conventional tea garden to a bio organic one, without any / much set back in productivity of tea.

Approximately one third of world food production is lost due to pests. If it were not the pesticides another one third would be lost. Nobel laureate Borlaug said in 1979 'we have to choose between feeding human beings or feeding pests'. A survey by National Council of Applied Economic Research has revealed that 'Every one rupee spent in plant protection would protect crop output worth rupees four , which would have been lost to pests'. While it is true that to increase crop yield pesticide has a great role to play but the application needs to be judicious . Excessive use of pesticides has become hazardous to human health as reported by many authors

Heavy use of chemical pesticides has taken serious turn in Thailand on the 'Hmong Farmers' health . It was observed that in Laguna and Nueva Eeija area of Philippines where the farmer use monocrotophos had health problem twice as compared to the farmers of Lucban and Quezon province who rarely used pesticides. It was reported that in 1989-90 more than 665 people died in India because of Pesticide poisoning and thousands of cases of crippling from pesticide have been recorded. With the growing consumption of pesticides the resistance of the insects against pesticides also increased. In 1934 there were 10 insects were known to be resistant against pesticides whereas by 1980 the figure had risen to 432. Pesticides were also found to affect the population of *Azotobacter* , *Rhizobium* etc. It reduces the number of nodules and affects mycorrhizal symbiosis. Owing to excessive use of pesticides alternation in the nitrogen balance occurs because pesticide affects nitrifying bacteria viz. *Nitrosomonas and Nitrobacter*. The list of adverse effect of synthetic pesticides in quite long and this emphasizes the need to search for other alternative method which would be effective in controlling the pests which causes economic injury to the crop and at the same time reduces the risk of health hazards.

There should be two distinct sets of practices in a bio-organic tea garden : i) Biological Pest Control ii) Biological Nutrient Management .- delete

DEMERITS OF SYNTHETIC PESTICIDES

Most of the pesticide residues affect the central nervous system, respiratory system and gastro-intestinal system of human beings

Depression, insomnia, hyper reflexia of human being are also caused by pesticides

The chemicals affect ground water resources through seepage

MERITS OF BIO ORGANIC CULTURE

It does not pollute the soil and ground water with chemical residues

It increases the biological diversity of plants and animal

It prevents leaching of minerals in the soil

It utilizes natural equilibrium for crop protection

DO NOT JUMP TO BIO ORGANIC OVERNIGHT

There can be a considerable reduction in productivity , if suddenly a tea garden switches over to bio organic form from a conventional form. There can be two distinct methods for such conversion :

1. Transform to bio organic through the pathways of Integrated Pest Management (IPM) and Integrated Nutrient Management (INM).

2. Convert one third of the tea garden into bio organic and then proceed in phasic manner.

INTEGRATION OF VARIOUS METHODS OF PEST MANAGEMENT SYSTEMS

1. PLANT RESISTANCE (VARIETAL CONTROL)

(Genetic make-up, tolerance use of mixed varieties)

In this case the resistance and susceptibility of a tea clone or seed stock in respect of any pest is taken into consideration. For example TV 23, Phoobshering 312 are susceptible to red spider mites; Teenali 17/1/54 is somewhat resistant to red spider mites. Hence in an area where infestation of mites is a problem better to avoid planting of TV 23 and Phoobshering 312. Tocklai Vegetative 1 (TV 1) is susceptible to Tea mosquito bug, hence where Tea mosquito bug infestation is prevalent, it is better not to plant TV 1. Tocklai Vegetative 9 (TV 9) is resistant to Violet root rot disease caused by *Spherostilbe repens*, hence for the waterlogged sections, planting of TV 9 should be considered as a good clone.

2. CULTURAL CONTROL

(Destruction of alternate hosts, barrier crop, raising of shade trees, irrigation & fertilizer practice, mulching.)

Most of the insects / mites need alternate host as weeds for completion of their life cycles, hence weed control is one of the best cultural control to keep pest population under control. In case of red spider mites, dust deposition on the leaves serves as the breeding ground for the mites, hence raising barrier hedge like *Adhatoda vasica, Duranta plumeiri* and *Lantana camara* can be planted on the boundary to prevent entry of dust. Even a row of tea plants along the roads can be left un plucked to take the shape of a hedge to prevent dust entry.

There are two categories of pests found in the tea garden. One is positively thermo (temperature) and photo (light) tropic, which means, if the light and temperature are more , these kind of organisms increases its population massively due to congenial weather conditions example red spider mites, thrips etc. Another type is negatively photo and thermo tropic, which means its population will increase where temperature is low and solar radiation is less example Tea mosquito bug, *Exobasidium vexans* (the causal organism of blister blight) etc. Hence, in case of the first type, adequate shade trees, which retains maximum foliage during dry period / drought should be there. There should be frequent irrigation to create a micro climate in the phyllospheric region to control pests.

In young tea, mulching i.e. covering of the barren surface with green mulches like water hyacinth, Guatemala grasses , thatch etc can be useful in maintaining a congenial agro-ecological condition, which will keep the pests under control.

3. BIOLOGICAL CONTROL

(Predators, parasitoids, pathogens, microbial pesticides e.g. Bacillus thuringiensis; Botanical antifeedant & repellants.)

Microbial pesticides, both entomo-pathogens and phyto - pathogens are now easily available in the market. Use of those against selective pests can be useful in controlling pests.

Plant extracts play a great role in controlling pests. Many weeds are used now-a-days for extracting juices, which act as anti feedant and repellant.

4. INTERFERENCE METHOD

(Pheromone traps, light traps, Sterile male Technique, Insect Growth Regulators e.g. diflubenzuron)

Although Interference method is still not very popular, it has considerable role in controlling pests. Different types of pheromone traps and light traps are available in the market now-a-days and many tea gardens have started using these devices. Insect growth regulators are also manufactured by many companies.

5. CHEMICAL CONTROL

(Chemical pesticides)

An elaborate list of chemical pesticides have been included in this book, with the chemical names, trade name, name of the manufacturing company, doses and target pests which can be used by the tea planters as a ready reckoner.

APPROPRIATE MANAGEMENT OF PEST CONTROL TECHNIQUES

Pest control programme will only be successful if appropriate management is done. It is necessary to identify the pests before applying any pesticide. Knowledge on the location of the particular pest in a plant , their life cycles, their damaging stages and period is of utmost importance. Spray techniques and spray schedule can only be properly prepared if the planter has a thorough knowledge about a pest. For examples are given below:

TARGET PESTS

1. To control thrips , jassids and scarlet mites, it is necessary to spray the under surface of the leaf, because these insects and mites remain on the ventral surface of the leaves.
2. To control tea mosquito bug, it is necessary to spray during early morning and evening because during heavy sunshine tea mosquito bugs escapes to the drains and weeds. It is also necessary to go for

barrier spray in case of tea mosquito bug. Barrier spray means, the boundary should be sprayed first using any fumigant so that the pest does not migrate to the adjacent section of tea. Tea mosquito bug is very fast in migrating.

3. To control looper caterpillars, it is necessary to spray the shade trees which serves as the alternate hosts of the caterpillars using foot sprayers or Bak Pak sprayers with BAN nozzle.
4. To control cockchafer grub, it is better to spray during evening because, they are more active during evenings.

PROPER PESTICIDE

Right kind of pesticides should be used to control pests. In case of wrong selection of pesticide, there will not only be wastage of money due to pesticide but it will also inflict crop loss due to delay in controlling.

PEST MAPPING & PEST SURVEILLANCE

Each tea garden should prepare a pest map for each tea section. The history of the section as far as pest infestation is concerned should be thoroughly studies. Data on agro-ecological condition, soil features, physiography, shade status, planting material, pest infestation type and time should be collected to prepare a section –wise pest map. This map will help the tea estates to plan their replanting programme, pest control strategy etc.

Every week a thorough study of the infestation status of pest in various section should be studied and a Pest Surveillance Weekly chart should be prepared to draw pesticide spray programme. This will serve as a radar for preparing pest control methodology.

CORRECT CHEMICAL & CORRECT DOSES

Right kind of pesticides should be chosen according to the species of the pest and intensity of infestation. Doses of the pesticides should be appropriate so that it reaches lethal dose, otherwise it may not be effective. Under dose can be very dangerous and it can increase the immunity of the pests.

There should be two distinct sets of practices in a bio-organic tea garden : i) Biological Pest Control ii) Biological Nutrient Management .

BIOLOGICAL PEST CONTROL PROTOCOLS IN COMPLETE BIO ORGANIC SECTIONS

CONTROL OF RED SPIDER & OTHER MITES

CULTURAL CONTROL

REGULATED SHADE : Mites are positively thermo and phototropic organisms. Too much infiltration of solar radiation and too much increase in temperature induce population explosion of mites. Hence it is necessary to cut off detrimental solar insolation and create a micro climate in the phyllospheric region conducive for the plants, by planting suitable shade trees. The species of shade trees should be selected taking into account the leaf fall period. Almost all the shade trees are leguminous , hence deciduous and shed leaves once in a year. It has to be seen that the leaf fall period of the chosen shade tree should not coincide with droughty period of the garden.

Shade trees have other important functions too:

i. Loppings, fallen leaves and twigs serve as humus

ii. Most of the shade trees being leguminous, have Nitrogen fixing capacity

iii. Creates a micro climate which provides soothing effect to the workers. Pluckers' productivity and workers' efficiency increase significantly under regulated shade

iv. Strong root system of the shade trees, breaks hard pan in the soil and paves the way for tea roots to proliferate. It also increases aeration inside the pore spaces

v. Shade trees act as wind barrier and protects the plantation from damage

vi. Thinning outs and loppings serve as fuel wood

vii. Controls many pests by buffering temperature

TOTAL CONTROL OF WEEDS : Weeds serve as alternate hosts for most of the insects and mites. These organisms (pests) complete their life cycles in two different species of plants. If weed is controlled, life cycle of the organisms is disturbed and reduction in population occurs.

PREVENTION OF DUST : Dust deposition on the leaves creates congenial substratum for laying of eggs of the mites. Hence it is necessary to prevent dust, especially in the roadside blocks, by raising boundary hedges.A row of tea bushes can be left unplucked after pruning or hedge plants like *Adhatoda vasica* or *Duranta plumeiri* or *Lantana camera* can be grown.

DEPLOYMENT OF WORKERS : Workers often serves as vectors (carrier) of mites. Through their bodies, plucking baskets etc mites are transmitted to unaffected sections. Hence deployment of workers should be from uninfested sections to the infested sections. Cattle also often serve as vectors. Hence, entry of cattle should be prevented.

IRRIGATION : When the temperature increases considerably and moisture stress

(vapour pressure deficit) is high, it is necessary to go for irrigation to induce humidity. Irrigation lowers the temperature thereby reduces the chance of mite infestation.

BIOLOGICAL CONTROL

DURING JAN - FEB

Spray at least two rounds of Sulphur – Neem combination with thorough drenching of bushes in the skiffed sections. BIO PESTICIDES

Paecilomyces fumosoroseus and *Verticillluim lecanii* in alternative rounds can be sprayed when population build up is nominal. Once , it is allowed to cross Economic Threshold Level (ETL), it will be difficult to control mites biologically. So, Pest Surveillance Technique is of immense importance in monitoring pest build up. It should be implemented meticulously to nip the pests in the bud stage.

CONTROL OF SUCKING PESTS

(Tea mosquito Bugs, Thrips, Jassids, Aphids etc)

CULTURAL CONTROL

- To control sucking pests , it is necessary to keep the plantation weed free.
- To control thrips, it is necessary to give a caustic /lime wash to the bush frame after pruning in the following ratio: Lime 1 kg + Caustic soda or washing soda @ 3 kg in 50 litres of water
- To control tea mosquito bugs, black plucking, slashing should be done if required.

BIOLOGICAL CONTROL

Spray *Beauveria* bassiana. While spraying against tea mosquito bugs, it would be necessary to go for barrier spray.

CONTROL OF CATERPILLARS

CULTURAL CONTROL

To control caterpillars , it is necessary to hand collect the pupae of the caterpillars from the bush and soil.

BIOLOGICAL CONTROL

Bacillus thuringiensis and *Nuclear Polyhedrosis Virus (NPV).* In case of red slug caterpillars, mid hamper of the bushes should be targeted.

OTHER PESTS

To control nematodes, ground application of *Paecilomyces lilacinus,* to control cockchafer and termites, *Metarhizium anisopliae* can be useful.

To control diseases caused by fungus and algae, spray of *Trichoderma viride* and *Pseudomonas flurescence* can be very useful. Cu- oxychloride is also allowed in organic tea gardens to control fungal and algal pathogens.

TO CONTROL RED RUST

Red rust disease is a secondary disease. Pre disposing factors for Red rust attack are i) Low Potassium content in the soil ii) Too dry condition and iii) waterlogged condition. Hence red rust disease which is caused by *Cephaleuros parasiticus* can be controlled by regulated shade, higher dose of Potassium and deepening of drains.

SPRAY OF BIO-CONTROL AGENTS

Spray while there is sufficient moisture in the atmosphere & soil

Use a sticker for better result

Spray while there is minimal sunshine

Wash the spray machines and containers before spray

There should not be any spray or broadcast of fungicide or bactericide before & after seven days' of spray .

POST PRUNING

To control *Poria* attack, immediately after , pruning, apply *Trichoderma* paste on the bigger cut marks and spray *Trichoderma* formulation@ 1 ml or 5 g per litre on the bush frame.

SPRAY OF BIO-CONTROL AGENTS

Spray while there is sufficient moisture in the atmosphere & soil

Use a sticker for better result

Spray while there is minimal sunshine

Wash the spray machines and containers before spray

There should not be any spray or broadcast of fungicide or bactericide before & after seven days' of spray .

ADDITION OF NEEM FORMULATION: With the bio pesticides, if neem product (*Azadirachta indica*) at recommended doses is added , efficacy can go up quite considerably and control may be prolonged.

POST PRUNING

To control *Poria* attack, immediately after , pruning, apply *Trichoderma* paste on the bigger cut marks and spray *Trichoderma* formulation @ 1 ml or 5 g per litre on the bush frame.

NATURAL ENEMIES OF TEA PESTS

Various predators, parasitoids and pathogens were found to be effective in controlling tea pests. Tea scientists have earmarked few predators/parasitoids/ pathogens as natural enemies of tea pests which are listed below:

Insect/Mite/Disease/Nematode	Predator/Parasitoids/Pathogen
Black Scale Red Spider Mite	*Chilocorus circumdatus* *Stethorus gilvifrons sp.* *Chrysoperla carnea* *Scolothrips sexmaculatus* *Paecilomyces fumosoroseus* *Verticillium lecanii*
Scarlet Mite	*Pronematus sp.* *Exothoris caudate* *Cunaxa sp* *Agisternus sp.* *Paecilomyces fumosoroseus* *Verticillium leccanii*
Pink Mite	*Amblyseius herbicolus* *Paecilomyces fumosoroseus* *Verticillium leccanii*
Tea Mosquito Bug	*Melamphus sp.* *Hexanemus sp.* *Oxyopes javanus* *Beauveria bassinana*
Looper Caterpillar	*Sarcophaga sp.* *Apantales sp.* *Argyrophylax sp* *Cotesia sp.* *Bacillus thuringiensis*
Red Slug Caterpillar	*Exorista sp.* *Argyrophylax sp.* *Cotesia sp* *Bacillus thuringiensis*
Eelworm' Leaf rollers	*Pasteuria neutrans* *Aphanogamus monitae* *Ascogaster sp.*
Flush worm	*Sympiesis dolicogaster*
Green fly	*Dyrid cephalosporium* *Beauveria bassinana*
Bunch Caterpillar Red rot	*Bacillus thuringiensis* *Trichoderma harzianum* *Ganoderma pseudoferreum*
Black rot	*Trichoderma viride*

USE OF WEED EXTRACT

There are many weeds, the extracts of which can help in controlling tea pests. Few of them are described below:

Name of the Weeds	Local Name
Ageratum conyzoides	Ilame
Polygonum runcinatum	Ratnaulo
Artemisia vulgaris	Titepati
Urtica dioecia	Shisnoo
Eupatorium glandulosum	Ban mara
Equisetum arvense	Kurkure jhar
Note : Soxhlet Apparatus is used for extraction. Petroleum ether, hexane/ Methanol used as solvent.	

BIOLOGICAL NUTRIENT MANAGEMENT

There should be a combination of bacteria or fungus based fertilizers, which are commonly termed as bio-fertilizers viz. *Azotobacter, Azospirillum,* Phosphorus Solubilizing Bacteria (PSB) or Phosphorus Solubilizing Microorganisms (PSM) like *Bacillus megatherium,* Vescicular Arbuscular Mycorrhizae (VAM) Potash Bacteria (*Frateuria aurentia*) and organic manures like farm yard manure, vermicompost, dried cow dung etc.

ROLE OF LIQUID BIO FERTILIZERS IN VARIOUS CROPS

1. Increases crop yield
2. Reduces need of chemical N,P & K fertilizers
3. Benefits to the next crop too, due to its residual effect
4. Enhances soil health and soil fertility
5. Returns are economically viable
6. Besides N.P. & K nutrition, growth hormones secreted increases the crop productivity. Disease and pest occurrence reduced

ROLE OF BIOFERTILIZERS IN TEA

- Curtails the requirement of inorganic fertilizers.
- Increases availability of nutrients.
- Gives 20 – 50 kg of Nitrogen and 30 – 50 kg of Phosphate in a year per hectare.
- Liberates growth promoting hormones vitamins; helps to improve soil fertility.
- Suppresses incidence of pathogens; secretes antibiotics to control diseases.
- Increases tea yield by 30 – 50 %

MERITS OF BIO ORGANIC CULTURE

- It does not pollute the soil and ground water with chemical residues
- It increases the biological diversity of plants and animal
- It prevents leaching of minerals in the soil
- It utilizes natural equilibrium for crop protection

DO NOT JUMP TO BIO ORGANIC OVERNIGHT

There can be a considerable reduction in productivity , if suddenly a tea garden switches over to bio organic form from a conventional form. There can be two distinct methods for such conversion :

i. Transform to bio organic through the pathways of Integrated Pest Management (IPM) and Integrated Nutrient Management (INM).

ii. Convert one third of the tea garden into bio organic and then proceed in phasic manner.

SPRAY OF BIO-CONTROL AGENTS

- Spray while there is sufficient moisture in the atmosphere & soil.
- Use a sticker for better result.
- Spray while there is minimal sunshine.
- Wash the spray machines and containers before spray.
- There should not be any spray or broadcast of fungicide or bactericide before & after seven days' of spray.

POST PRUNING

To control *Poria* attack, immediately after , pruning, apply *Trichoderma* paste on the bigger cut marks and spray *Trichoderma* @ 1 ml /5 g per litre on the bush frame.

NUTRIENT MANAGEMENT

In a bio tea garden use of inorganic fertilizers is not allowed. Hence, organic manures and bio fertilizer combination should be used.

INPUTS IN PLANTING PITS

At the time of new planting / infilling following ingredients may be used :

Cow dung – 1kg + SSP 30 g + Oil Cakes – 30 g per pit .

INPUTS IN NEW PLANTATION AFTER PLANTING UPTO ONE YEAR

Application of cow dung lassi at monthly interval on the soil of the collar region can be quite useful during droughty periods.

Humik acid liquid can be applied on the soil of the collar region at least twice a year @ 2500 ml per hectare.

INPUTS AFTER PRUNING

All the pruning litters to be buried inside the soil by digging trench in the interspaces. Add Cowdung @ 1000 g per plant.

GENERAL MANURING

Since inorganic NPK will not be allowed in a bio tea garden, organic manures and bio fertilizers should be used extensively . Following proposition can be thought of :

Placement of cow dung @ 1 - 2 kg per plant, by digging a small pit in between two bushes (or by broadcast if availability of workers is a problem.

During rainy season, during rain free days, broadcast of bio fertilizers in the following proportions can be used:

Azotobacter @ 2.5 kg + Phosphorus solubilizing bacteria @ 5 kg + Vescicular arbuscular mycorrhizae @ 5 kg per hectare can be applied.

GREEN CROPPING

In the new plantation area, green crops and cover crops like *Crotalaria anagyroides*, *Tripsacum laxum*(Guatemala grass), *Mimosa invisa* etc should be planted to increase organic matter content of the soil.

NUTRITIONAL VALUES OF VARIOUS ORGANIC MATTERS

	N %	P %	K %
Farm Yard Manure	0.5	0.2	0.5
Tea pruning	2.5	0.5	1.5
Tea waste	3.6	0.6	2.3

OIL CAKES

Castor	**5.0**	**2.0**	**1.0**
Ground Nut	7.4	1.5	2.4
Neem	5.2	1.0	1.4
Lin seed	4.7	1.7	1.3
Coconut	3.4	1.5	2.0
Sesamum	5.5	2.1	1.3

LOPPINGS

Tripsacum	**1.6**	**0.6**	**1.7**
Napier	0.8	0.4	1.3
Mimosa invisa	2.6	0.6	1.0
Thatch grass	0.5	0.1	0.5

MULCHING

Tea prunings	2.5	0.5	1.5
Tea waste	3.6	0.6	2.3
Paddy straw	0.5	0.07	0.6
Thatch grass	0.5	0.1	0.5
Guatemala grass	1.6	0.4	1.3

AVENUE PLANTATION

On the boundary of each section, Neem, Karanj (*Pongamia*) and madar (*Erythrina indica*) type of plants should be planted at 10 m spacing. In between these trees , small herbs like *Leucaenea leucocephala* (Su babool) , *Sesbania* (dhaincha) should be planted at 1 m distance. These will add sufficient quantity of bio mass in the soil.

SHADE PLANTATION

Shade trees contribute about 15 tonnes of leaf litter per year . Hence leguminous shade trees should be planted in all the sections at recommended spacing. Shade tree loppings can provide nutrients as furnished below:

Nutrient	Kg/ha
Nitrogen	63 to 126
P_2O_5	18 – 36
K_2O	22-44
Ca	32-64
MgO	16-32

ROCK PHOSPHATE/WOOD ASH

Rock phosphate is allowed in a bio tea garden. P_2O_5 @ 50 kg per hectare per year in the form of Rock Phosphate can be applied. Wood ash is a good source of Potassium in organic form. Apart from using Potash bacteria - *Frateuria aurentia* @ 2.5 kg per hectare, wood ash can be used to replenish Potassium in the soil. Wood ash can be obtained by burning shade tree loppings etc. Dose can be ascertained by determining the Potassium content of the wood ash in a recognized laboratory.

WEED CONTROL

No herbicide can be used in a Bio Organic garden. Only manual weed control is allowed.

During winter, cheeling (cheel hoeing) operation should be undertaken in all the sections. In case of heavy weed infestation, chop hoeing can be done. In rainy seasons weeds should be controlled by sickling. Weed control is easy in skiffed sections where canopy itself prevents entry of sunlight to the soil. In Young tea sections, mulching should be done to control weeds.

Extensive research is going on to evolve biological control of weeds by using fungal preparations. When such bio control agents will be available, weed control biologically can be easier.

FOLIAR SPRAY

Bio Foliar spray of sea weed extract @ 500 ml per hectare can be safely used to provide nutrients to the plants.

Foliar spray of Humic Acid (or a combination of Humic acid and fulvic acid) @ 500 ml per hectare can be quite useful.

BIO DYNAMIC PREPARATION

Now a days bio dynamic preparations are gaining ground in the Bio-Organic tea gardens. Few examples are given below :

BIO DYNAMICS - 500 (Cow Horn Manure)

Cow dung is stuffed inside the cow horns and buried in a pit. The depth of the pit should be 45 cm. The pit is covered with good quality compost manure.

The distance between two cow horns should be at least 5 cm. The open end of the cow horns should be at the bottom so that no water accumulation is possible. The pit should be mulched by paddy straw and a thatch shade should be constructed on top of the buried cow horns. After about 4 months the dung inside the cow horn turns brown it emits a sweet smell. It should be then taken out of the soil and stored in a cool place and must be kept in glass jars surrounded by damp coir pith.

Dose : 100 gms of this brown cow dung should be sprayed on the tea foliage dissolving in 60 litres of water. Before sprayed it should be stirred vigorously

Time of spray : Post monsoon

BIO DYNAMICS 501 (Cow horn Silica):

In this method, instead of cow dung, finely ground quartz crystals of silica are inserted inside the cow horn. This preparation is done during hot summer months. Other methods are same as Biodynamics 500. The color of mature horn silica is grey or yellow. It is stored in a glass jar on a sunny window site. It can be stored for 2-3 years.

Dose : 2.5 gms per litre of water. Before sprayed it should be stirred vigorously

Time of spray : July to October.

BIO DYNAMICS 502: is made from the flowers of yarrow (*Achillea millifalium*). The dry flowers of yarrow are put in stag bladder until the bladder is packed, then stitch the slit with cotton thread. Store in a closed basket to keep away rodents or pests. Hang up in March to get cosmic influences till September. Burry it in September in a mud made flower pot surrounded with fertile soil. After 4-5 month it may be taken from the soil in February or March. It must be stored in cool, dark and moist place.

BIO DYNAMICS 503: is made from the flowers of German Chamomile (*Matricaria chamomile*). Cow or Bull small intestine is needed for this preparation. The intestine at first cut into 15cm bits and tie the cut bits at one end by cotton thread. Then put a funnel to the open end and fill with dry flowers of chamomile. The packing should not be to hard or loose. Packed intestine just looks like sausages. Stack the sausages into a bundle which could be placed in a mud pot which must be surrounded with fertile soil or old compost. Bury it in October and let it remain in the soil till Feb/March.

BIO DYNAMICS 504: is made from the leaves of Stinging nettle (*Urtica parviflora*). Fill up the dry leaves in mud pots. Press well into the containers; ensure that the lid is on. Place the pot under the influence of Mars. Place it into soil in early September and lift the preparation after one year.

BIO DYNAMICS 505: is made from the bark of the Oak (*Quercus glanca*) tree.The bark must be collected from minimum 25 years old oak tree. Finely crush oak bark put in cow skull. Block the opening with a well shaped bone pieces or stone. Place the skull in a watery environment. The preparation is placed in September and lifted in March.

BIO DYNAMICS 506: is made from the flowers of Dandelion (T*araxicum officinalis*). Place the dry flower in a mesentery of a cow. Ensure that the extra fat is cut off. Do not wash the mesentery and wrap it like a parcel and tie with a jute thread. Place the parcel in a good mixture of soil and compost into a pot. Place the preparation in September and Lift it in March. This preparation is related mainly with salicylic acid. It helps to draw energy from the planet.

BIO DYNAMICS 507: is made from the flowers of Valerian (*Valeriana officinalis*). It is the liquid preparation among all the Biodynamic preparation. The clipped flowers put into a mortar and grind to form a paste. This paste is added to water in the ratio of **1:4** in a bottle. Store in a cool place. It contains **5%** solution.

Biodynamics 502 to 507 are used in compost making

INSECT GROWTH REGULATORS

These chemicals interfere with the secretion of endocrine glands and thus interfere with chitin synthesis, an essential component of cuticle e.g. Diflubenzuron

SEX PHEROMONES

Now a days, many types of sex pheromones are being experimented at many tea research centres to ascertain its efficacy in controlling tea pests.

DIFFERENT BIO CONTROL AGENTS (ENTOMO PATHOGENS & PHYTO PATHOGENS)

Paecilomyces fumosoroseus : Controls mites including red spider mites

Characteristic features : Conidia is long ovoid in shape. Pink grey , pin tan, tan, grey, reverse pale to yellow in colour. Conidiophore is smooth and colourless; conidial chain long and conidial head are diffuse.

Mode of Action : The spores of *Paecilomyces fumosoroseus* comes in contact with the pests, germinates , multiplies and penetrate into the haemocoel. It takes nutrient from the haemolymph and colonize the whole visceral mass.

Paecilomyces lilacinus : Controls nematodes

Characteristic features : Conidia is short fusoid, lemon shaped. Purple grey, tan or grey in colour. Conidiophore often roughened and slightly coloured; conidial heads compact with short conidial chains.

Mode of Action : *Paecilomyces lilacinus* produces hyphae, which infects nematodes like *Meloidogyne incognita, Rotylenchulus reniformis, Helicotylenchus incisus, Pratylenchus coffeae* etc. *Paecilomyces* is a facultative parasite of root knot nematode eggs. It disrupts the lip and chitin layer of egg shell.

Verticillium lecanii:

Controls mites, aphids,thrips and scale insects.

Characteristic features : Conidia cylindrical with rounded ends. Conidiogenus cells narrow, awl-like, tapering towards tip.

Mode of Action : *Verticillium* enters the body of the insects through insect cuticle and sporulates inside to produce a toxin called bassianolide.

Beauveria bassiana:

Controls sucking pests like flush worms, tea mosquito bugs etc. It can also control looper caterpillars to some extent.

Characteristic features : Conidia globose or sub globose in shape. Pink-grey, pink-tan, grey in colour. Conidiophore smooth and colourless. Conidial chains often long and conidial heads diffuse.

Mode of Action : The spore of *Beauveria bassiana* germinates while it comes into contact with pests. It penetrates through cuticle and multiply inside. It produces a toxic metabolite called Beauvericin. This product inhibits synthesis of metabolic enzymes and leads to paralysis of physiological functions.

Metarhizium anisopliae: Control soil borne insects like termite, cockchafer etc.It can also control pupae of looper caterpillars.

Characteristic features : Conidia cylindrical with middle constricted, green, yellow-green or brown coloured. More often affecting hosts in soil rather than aerial parts.

Mode of Action : *Metarhizium* produces a number of secondary metabolites acting as mycotoxins like Destruxin E. These toxins cause titanic paralysis and rapid development of mycotoxins causes death of termites.

Trichoderma viride :Antifungal/ cures root diseases ; revives plants suffering under waterlogged condition

Mode of Action : The conidia of *Trichoderma viride* come into contact with target host. It germinates and the germ tube coils the hyphae of pathogenic fungi, degrades the cell wall and feeds on the nutrients causing mycoparasitism. It also produce toxic metabolites.

Pseudomonas fluorescence: Against diseases

Mode of Action : *Pseudomonas fluorescens* controls diseases by producing antibiotics and also obstruct supply of food to the pathogenic fungi.

Bacillus thuringiensis: Used against looper and other caterpillars.

Mode of Action : B. thuringiensis produce a class of insecticidal small molecules called beta-exotoxin, the common name for which is thuringiensin. Beta-exotoxin and the other Bacillus toxins contribute to the insecticidal toxicity of the bacterium to lepidopteran.

CALENDAR OF METHODOLOGY FOR A BIO-ORGANIC TEA GARDEN

JANUARY

FOLIAR SPRAY

One round of Neem and Sulphur combination as a prophylactic spray against mites and sucking pests. Neem @ 1000 ml + Sulphur @ 2.5 kg per hectare

FEBRUARY

FOLIAR SPRAY (BIO PESTICIDES)

Paecilomyces fumosoroseus @ 250 ml + Neem @ 1000 ml per hectare

MARCH

FOLIAR SPRAY (NUTRIENTS & GROWTH PROMOTERS)

Sea Weed Extract @ 500 ml + Humik Acid @ 500 ml per hectare

FOLIAR SPRAY (BIO PESTICIDES)

Verticillium lecanii 250 ml + *Paecilomyces fumosoroseus* @ 250 ml per hectare

APRIL

FOLIAR SPRAY (BIO PESTICIDES)

Beauveria bassiana @ 100 ml + Neem @ 1000 ml per hectare

MAY

FOLIAR SPRAY (NUTRIENTS & GROWTH PROMOTERS)

Sea Weed Extract @ 500 ml + Humik acid @ 500 ml per hectare

GROUND APPLICATION (NUTRIENTS)

Placement of Cowdung @1.5 tonnes + Vermicasting @ 30 kg *Trichoderma* @ 2.5 kg (light watering/irrigation after application if sufficient moisture is not present in the soil)

JUNE- AUGUST (MONSOON)

Need based application.

FOLIAR SPRAY (BIO PESTICIDES)

Beauveria bassiana @ 100 ml + Neem @ 1000 ml per hectare once in each month

FOLIAR SPRAY (BIO PESTICIDES)

Pseudomonas fluorescence @ 250 ml per hectare

SEPTEMBER

FOLIAR SPRAY (NUTRIENTS & GROWTH PROMOTERS)

Sea weed extract@ 500 ml + Humic acid formulation@ 500 ml per hectare

GROUND APPLICATION (NUTRIENTS)

Placement of Cowdung @1.5 tonnes + Vermicasting @ 30 kg *Trichoderma* @ 2.5 kg (light watering/irrigation after application if sufficient moisture is not present in the soil)

OCTOBER

FOLIAR SPRAY (BIO PESTICIDES)

Paecilomyces fumosoroseus @ 250 ml + Neem @ 1000 ml per hectare

NOVEMBER – DECEMBER

No application

Note :

i. In case of looper attack use *Bacillus thuringiensis* at recommended doses as and when required.

ii. Bio control agents, mainly entomopathogens should be used only if pest is present.

iii. Biodynamic formulations can be applied as foliar nutrients

PEST SURVEILLANCE

"Prevention is better than cure", "A stitch in time saves Nine" and " Nip in the bud"- these three proverbs are very appropriate in tea gardens for developing a well-knit pest control mechanism. Before taking out the guns to eradicate pest problems, it is pertinent to know the kind of pest attack and its intensity through effective feed back system. Like a successful warrior, it is important to get a clear picture about the strength of the enemy and then formulate a full-proof war strategy. In a tea garden, there should be a method to regularly study the intensity of pest attack in each tea section through PEST SURVEILLANCE technique. An Assistant Manager in a tea garden can be given the responsibility to visit each and every section once a week and prepare a Pest Surveillance color chart. A design of the chart is furnished below as ready reference:

PEST SURVEILLANCE

Week :

Section #	Year of Planting	Area (ha)	Red spider mites	Thrips	Jassids	Tea Mosquito Bug	Caterpillars	Cockchafer	Red rust	Other pests

RED Severe infestation, when immediate spray is required

YELLOW Considerable infestation but spray of pesticide may be deferred.

BLUE Moderate infestation, to be decided if pesticide spray is at all required or not.

GREEN Pesticide spray not required.

From this chart, it will be easy to prepare pesticide spray schedule. The 'red' marked sections should be sprayed first and then the 'yellow' sections. In case of 'red' marked sections also the priority should be given on the basis of the species of the insect / mite and causal organism of fungus. If the section is infested by looper caterpillars or red slug caterpillars, it should get top priority rather than red spider mites or scarlet mites.

COMPOST PIT

- Construct a tank of the size : length 3 metres, width 2 metres and height 1.5 metres with pigeon holes.
- Put a plaster of liquid cowdung inside the tank like plastering huts.
- Spread a polythene sheet on the floor.
- Spread green leaves, agro waste etc about 125 – 150 Kg to the height of 6 inches.
- Mix about 10 Kg dung in 100 litres water and spray on the greeneries.
- Spread about 125 – 150 Kg of cattle urinated soil on top of it.
- This process to be continued till the tank is filled.
- Cover the tank with polythene sheet.
- After 10-15 days fill the gap in the same manner.
- Within 40 – 50 days compost will be ready.

Note: one tank can produce 12 M.T. of organic manure in a year

VERMICULTURE

WORMS USED

- *Eisenia foetida*
- *Amyanthas diffringens*
- *Eudrilus eugeniae*
- *Perionyx excavatus*
- *Lampito mauritii*
- *Pheretima elongate*
- *Pontoscolex corethrurus*

NUTRIENT CONTENT OF VERMICOMPOST

Nutrients	Content (approximately)
N (%)	GREEN Pesticide spray not required.
NH4-N (ppm)	85
NO3-N (ppm)	100
P2O5 (%)	0.75 – 1.12
K2O (%)	0.65 – 1.14
S (%)	0.24-0.52
pH	6.76 – 7.28
C/N Ratio	12
Humic acid	5
CEC(cmol (p+)kg-1	95
Zn (ppm)	55

REARING OF WORMS

- Wooden box of 45 cm X 65 cm X 20 cm with drainage holes.
- Rice husks, cereal straw, saw dust are used as breeding material.
- Ferment for one month.
- Mix Guatemala grass / water hyacinth / poultry manure.
- pH should be neutral, temperature 20 – 27 ^{0}C .
- 100 g of worms from breeder's stock required for each 0.1 m ^{2} of surface area.

COMPOSTING

- Pit size – 2 m x 1 m x 1 m.
- Bricks and sand at the base.
- Line the pit with old gunny bags.
- Fill the pit with FYM, green weeds and other organic residues.
- A layer of 15 cm. Loamy soil to be sprinkled and keep moist.
- Two third of the worms from rearing box to be harvested and placed on top of the heap.
- 10 kg. of vermicasting is produced per kg. of worm.

CONTROL MEASURES

Bio-pesticide: *Metarhizium anisoplae*: 800 g./ drum

Synthetic chemicals:

1. Imidacloprid: 80-100 ml./drum
2. Chlorpyriphos :300ml./drum

Termite

Species of Termite:

1. Live wood eating termites: *Microcerotermes* sp.
2. Scavenging termites: *Odontotermes assamensis*

Seasonal Occurrence:

Initial infestation starts:– August – October

Peak:- November – February (Winter)

Site of attack: Trunk, branches and stems.

Kinds of Termites:

a. Live wood eating termites form many small nests of 8 – 20 cm. below in the soil.
b. Queens of termites produce eggs every second, hence must be killed for effective control.

DAMAGE

1. The scavenger first attack the dead wood resulting from fungal or borer attack, also extended the damage to the adjacent healthy tissues by removing the dead woods.
2. The live wood-eating termites attack the tissue of the bark of healthy young bushes. They excavate fine tunnels into the healthy wood and hollow out the stem leaving only a thin layer of wood and bark or whole woody part with earth materials leaving the bark out side.

CONTROL MEASURES

Cultural

i. Clean the run in the infected bush thoroughly at pruning and put in-dopaste at the cut ends.
ii. Improve drainage and shade status.
iii. Destroy termite mound and queens.

Bio-Pesticide:

Metarhizium anisoplae : Dose: 800 g./ drum

Exercise No. 10

Tea Pests and Pest Control

DEFINITION OF PESTS

1. Pests are any form of plant or animal life that is injurious or potentially injurious to human being and its economy.
2. A plant or animal detrimental to humans or human concerns (as agriculture or livestock production)
3. Example: insects, vertebrates, weeds, pathogens such as bacteria, fungi, viruses.

INSECTS

1. Usually small arthropod animals of the class Insecta,
2. Having an adult stage characterized by three pairs of legs and a body segmented into head, thorax, and abdomen and usually having two pairs of wings.
3. Insects include the flies, crickets, mosquitoes, beetles, butterflies and bees.

MITES

1. A tiny four or eight-legged creature belonging to the Order Acarina related to spiders.
2. Some mites live freely and others as parasites. Mites can attack plants and animals, carry disease, and cause allergies.

ROOT DAMAGING PESTS

ROOT KNOT NEMATODES (EEL WORMS)

Root Knot Nematodes are commonly known as eelworm (Order: Nematode).

The pathogenic nematodes are of two types:

1. Root-Knot Nematodes,
2. Root-lesion nematodes

Harmful species. of Root-Knot Nematodes are:

a. a) Meloidogyne incognita
b. b) Meloidogyne javanica
c. c) Meloidogyne hapla
d. d) Meloidogyne brevicaudata

Harmful species of Root-lesion Nematodes are:

i. *Pratylenchus brachyurus*
ii. *Pratylenchus loosi*

DAMAGE SYMPTOMS OF NEMATODES

Root-Knot Nematodes

1. Usually attack young seedlings and clonal plants in nursery and sometimes in the field during the first year of planting.
2. The infected roots exhibit numerous galls or knots, swelling, curved and twisted formations .
3. These changes interfere with the uptake of water and nutrients as a result plants suffer from drought.

Root Lesion Nematodes:

1. They feed and lay eggs within the cortex of young roots and move from root to root in search of food.
2. These movements and feeding activity disintegrate cortical tissues and retard growth of the plant.
3. Such debilitated plants may not recover after pruning.

Few Facts and Control Measuresa

1. Care should be taken in the selection of nursery sites since nurseries play an important role in the spread of nematodes.
2. Virgin land or land under wild vegetation for a number of years should be selected for nursery.
3. Soil considered unsuitable for nursery on account of high eelworm count can be rehabilitated by putting under green crop *Crotolaria anagyroides* at least for two years. *Crotolaria* is not infested by eelworm.
4. It should not be above 4 nos./10 g. of soil. If it is more it can be controlled by the following methods :

a. Calculate the quantity of soil required to fill 1000 polythene sleeves. In that quantity of soil, add 1000 g of carbofuran thoroughly or 2 g. carbofuran / sleeve,
b. Uniform heating (after sieving twice through GI wire mesh No. 4 for eliminating any vegetative matter) of the soil up to 60-70^{0}c for 4-5 minutes on plain tin sheets.

COCKCHAFER GRUB

Cockchafer Beetle- Adult

Out of several species of cockchafer grub particularly *Holotrichia impressa* (Order: Coleoptera) is the most destructive in Dooars and Terai.

Identification

Eggs are oblong and glistering white in colour,

Grub whitish in colour,

Pupae yellowish in colour (20-22 mm in length),

Adult beetles are deep brown in colour.

Site of Attack

a. It generally found in nursery as well as in field up to 3 years after transplanting
b. The grub attacks collar region of young plant (below 3 years) and seed-ling & roots of young tea,
c. The adult cause damage to foliage of shade trees.

Damage Symptoms

1. Larvae eat away the bark in the collar region just below the soil surfac either in a ring or in patches.
2. The damage done by cockchafer is very similar to that of manur damage.

Life Cycle

Site of Oviposition:

Eggs are laid in the soil at a depth of 4-5 cm.

Fecundity: 40 eggs

Incubation Period: 10-11 days

Peak period of infestation: June – August (grub) During April beetles s down on the trunk of shade trees, fencing etc. where from these can be h collected and destroyed.

Synthetic Chemicals:

Imidacloprid: 80 – 100 ml./ drum

Chlorpyriphos (Dursban): 300 ml./ drum

1. **Damaging Stage**:- Larva

2. **Adult**

Orange coloured, medium sized moth.

The adult has white and black spotted wings.

The larva is red coloured.

3. Life Cycle

Eggs are usually laid singly on the bark.

Eggs hatch in 8 – 12 days and larval development takes 3 – 4 months.

Pupation takes place inside the stem and the adult emerges in about 2 months.

Adult is a moth.

DAMAGE SYMPTOMS

Caterpillar bores into the stem or branches to feeds on the wood.

In early stages of attack, young plants or braches show wilting.

Pellet-like excrement of the larva hangs out and accumulates at the base of the plant.

In advanced cases, the branch dries up.

CONTROL MEASURES

Affected plants or twigs should be cut and burnt.

Providing good shade to plantation.

Field release of white muscardine fungus *Beauveria bassiana* and parasite *Amyosoma zeuzerae.*

STEM BORER Xyleborus Fornicates (Order: Coleoptera)

Damaging stage : Adult

Life Cycle

Incubation- 10-12 days

Larval period- 45-60days

Pupal period- 20-30 days (tunnel)

Activity- Winter months

MAJOR SUCKING PESTS

1. Tea Mosquito bug : *Helopeltis theivora*
2. Tea Thrips : *Scirtothrips dorsalis*
3. Tea jassids : *Empoasca lavescens*
4. Aphis : *Toxoptera aurantii*

TEA MOSQUITO BUG

Leaf Crippled After Severe Attack of Tea Mosquito Bug

Damaging stage – Nymphs and Adults

Seasonal Incidence

i. Feb – Mar:- Building up of the population
ii. Apr – May:- Infesting tea leaves
iii. June – sept:- attaining peak
iv. Oct – Mid Dec:- Retaining of population
v. January:- dwindles down gradually

Site of Attack

The nymphs and adults suck the sap of young leaves, buds and tender stems,

Growing shoots and bud at the leaf axils are the oviposition sites.

LIFE CYCLE OF TMB

Adult lays eggs on midrib of leaves

Fecundity - A bug may lay 4-5 eggs/day. A single female may lay more than 200 eggs.

Incubation period- 10-13 days

Life cycle of *Helopeltis* completed in 10-35 days.

Nymphal period- 22-23 days

Adult - Black head, pale yellow and black thorax, yellow and greenish black abdomen.

Nymph is dirty yellow

DAMAGE SYMPTOMS

1. Tea leaves are damaged by the nymphs and adults.
2. Initially, a ring like spot is formed at the sap-sucking point turning translucent light brown within 24 hours.
3. These spots later on turn dark brown and appear as sunken spots and dry up.
4. An adult can produce about 100 such spots within 24 hours.

TEA THRIPS *Scirtothrips dorsalis* (Order: Hemiptera)

Damaging stage – Nymphs and Adults

Seasonal incidence: March – April: Rapid Build up

May: Attaining peak

Life cycle: Fecundity: Eggs are laid singly, completely embedded in the tissue of leaf buds and young leaves

Incubation period: 3 – 4 days

Larva: 1 – 3 days

Pupa: 3 days. Pupation takes place in lichens and mosses growing on tea bushes.

Total duration of life cycle: 14 – 20 days.

DAMAGE SYMPTOMS

1. Leaf surface becomes uneven, curly and matty, exhibiting parallel lines of feeding marks on either side of the midrib. Puncture marks appear as minute brown spots, later on 'Sand papery' lines (2 - 4) appear on either side of the midrib.
2. Heavily infested fields sometimes acquire a bronze colour.
3. Badly damaged leaves appear rough, deformed and curled.

TEA JASSIDS : Empoasca Flavescens (Order : Hemiptera)

Damaging Stage- Nymphs and Adults

Seasonal incidence:

a. May – July: The pest is more active
b. August onward: Population suddenly declines to a negligible number.

Life Cycle

1. Fecundity: 100 eggs during her lifetime.
2. Depending on the temperature, the eggs will hatch after 6-13 days.
3. Eggs are laid by inserting in the tissue of new tea buds and shoot, particularly in the internode. In one bud up to 7 eggs are laid.
4. Nymphal period: Total nymphal development period is 7 - 16 days being faster in warm weather condition. There are five nymph stages.
5. Total duration : The life cycle completed in 14 – 28 during March-July.

DAMAGE SYMPTOMS

1. Both adult and nymphs, suck the sap of young leaves, cause damage, nymphs are responsible for greater damage than the adults.
2. The affected leaves become uneven ad the leaves curl downward, the margin become recurved and subsequently turns brown and dry up. This characteristic symptom is known as the rim blight.
3. The mid rib and veins of the affected leaves may also show somewhat brownish discolouration .

APHIDS – Toxoptera Aurantii (Order : Hemiptera)

Damaging stage- Nymphs and adults

Site of attack: Buds, Tender stems, Young leaves..

Seasonal incidence: January – April

Damage symptoms: Crinkled and curled leaves below tipping level.

CONTROL MEASURES

Cultural Control

1. Thinning out of shade trees in heavy shaded areas.
2. Check the egg population either by hard plucking or level of skiff in heavily infested section.
3. Prune/deep skiff heavily infested sections by *Helopeltis* in the previous season.
4. Remove weed or other alternate hosts. Alternate weed hosts for *Helopeltis* are *Melastoma malabathricum, Mikania micrantha, Oxalis acetocella.*
5. Alkaline/ lime/ caustic washing of the trunk of the bushes after cleaning the mosses and lichens and stirring of soil around the collar region will kill the pupae of thrips during Dec. - March. Lime 1 kg + Caustic soda or washing soda @ 3 kg in 50 litres of water.
6. Ladybeetles, praying mantids feed on jassids.

7. Tea growing under shade trees tends to have less jassid problem. Mulching also helps to increase humidity and therefore reduce jassid populations.
8. 8. Use of light taps.
9. 9. Sex-Pheromone traps for example: Sticky Delta Trap, Wota Trap, Fero Trap.

Mechanical and Physical Control:

1. Hand collection of the adults & nymphs of Tea Mosquito bug – the adults and nymphs are active during early morning, late afternoon and night on the top hamper of bushes.
2. Use of yellow colour sticky traps for monitoring of thrips.

Bio-Pesticide:

1. *Beauveria bassiana*:- Dose: (i) Liquid 200 ml./ drum
   (ii) Powder: 600 g./ drum
2. *Verticillium lecanii*:- Dose: (i) Liquid 200 ml./ drum
   (ii) Powder: 800 g./ drum

LEAF EATING PESTS

CHARACTERISTICS OF LEPIDOPTERAN INSECTS

Lepidopteran species are characterized by more than three derived features, some of the most apparent being the scales covering their bodies and wings, and a proboscis. The scales are modified, flattened "hairs", and give butterflies and moths their extraordinary variety of colours and patterns. Almost all species have some form of membranous wings, except for a few that have reduced wings or are wingless. Like most other insects, butterflies and moths are holometabolous, which means they undergo complete metamorphosis. Mating and the laying of eggs are carried out by adults, normally near or on host plants for the larvae. The larvae are commonly called caterpillars, and are completely different from their adult moth or butterfly form, having a cylindrical body with a well-developed head, mandible mouth parts, and from 0 to 11 (usually 8) pairs of prolegs. As they grow, these larvae will change in appearance, going through a series of stages called instars. Once fully matured, the larva develops into a pupa, referred to as a chrysalis in the case of butterflies and a cocoon in the case of moths.

EXTERNAL MORPHOLOGY

- Lepidoptera are morphologically distinguished from other orders principally by the presence of scales on the external parts of the body and appendages, especially the wings.

- Lepidopterans undergo a four-stage life cycle: egg; larva or caterpillar; pupa or chrysalis; and imago/ adult and show many variations of the basic body structure, which have evolved to gain advantages in life-style and distribution.
- Body is divided mainly into three parts - head, thorax and abdomen.

HEAD

The head is where many sensing organs and the mouth parts are found. Like the adult, the larva also have a toughened, or sclerotized head capsule. There are two compound eyes, and chaetosema, raised spots or clusters of sensory bristles unique to Lepidoptera, even though many taxa have lost one or both of these spots. The larvae, called caterpillars, have a toughened head capsule. Caterpillars lack the proboscis and have separate chewing mouthparts. These mouthparts, called mandibles, are used to chew up the plant matter that the larvae eat. The lower jaw, or labium. The head is made of large lateral lobes, each having an ellipse of up to six simple eyes.

THORAX

The thorax is made of three fused segments - the prothorax, mesothorax, and metathorax,each with a pair of legs. The first segment contains the first pair of legs. The three pairs of legs are covered with scales. Lepidoptera also have olfactory organs on their feet, which aid the butterfly in "tasting" or "smelling" out its food. The two pairs of wings are found on the middle and third segment, or mesothorax and metathorax respectively. The caterpillar has an elongated soft body that may have hair-like or other projections, 3 pairs of true legs, with 0–11 pairs of abdominal legs (usually 8) and hooklets, called apical crochets. The thorax will usually have a pair of legs on each segment. The thorax is also lined with many spiracles on both the mesothorax and metathorax, except for a few aquatic species, which instead have a form of gill.

ABDOMEN

The abdomen, which is less sclerotized than the thorax, consists of 10 segments with membranes in between allowing for articulated movement. The sternum, on the first segment, is small in some families and is completely absent in others. The last 2 or 3 segments form the external parts of the species' sex organs. The abdomen of the caterpillar has 4 pairs of prolegs, normally located on the third to sixth segments of the abdomen, and a separate pair of prolegs by the anus, which has a pair of tiny hooks called crotchets. These aid in gripping and walking, especially in species that lack many prolegs.

SCALES

The wings, head, parts of the thorax and abdomen of Lepidoptera are covered with minute scales, a feature from which the order 'Lepidoptera' derives its name. Most scales are lamellar, or blade-like and attached with a pedicel, while other forms may be hair-like or specialized as secondary sexual characteristics.

The lumen or surface of the lamella has a complex structure. It gives colour either by coloured pigents that it contains, or through structural coloration with mechanisms that include photonic crystals and diffraction gratings.

CATERPILLARS

Looper Caterpillar: *Biston (Buzura) suppressaria , Hyposidra talaca, Hyposidra infixaria, Ectropis sp. , Ascotis sp, Cleora sp.*

Red slug caterpillar: *Eterusia magnifica*

Bunch Caterpillar: *Andraca bipunctata*

LOOPER CATERPILLARS

1. Time of Occurrence:

a. Caterpillar: March - October
b. Pupa: April – November
c. Moth: Feb – Mar/May/Jul/Aug - Oct

2. Eggs and Life Cycle:

a. Bluish green covered with buff coloured hairs, laid on shade tree trunk.
b. Hatch in 8 – 9 days into black caterpillars.
c. Larval stage: 24 – 36 days
d. Pupation: 20 – 22 days in soil at 5 – 10 cm. Deep

DIFFERENT TYPES OF LOOPER CATERPILLARS

Biston (=Buzura) suppressaria

This is the earliest known species of looper caterpillar infesting tea. The moth is greyish white, finely speckled with black. The males are generally smaller than the females and are easily distinguished by the bipectinate (feathered) antennae. The eggs are laid in heaps, each containing 200-600 eggs covered with a buff coloured hairs. The most common site for the deposition of eggs is the trunk of shade trees or any other tall trees in the vicinity of tea area. The young caterpillar is dark brown with greenish white lines along the back and side. The body colour soon turns to light green and with age, it acquires a brownish grey colour similar to that of a mature twig of tea.

Hyposidra Talaca

Early instars caterpillars are brown to black in colour with transverse six white bands on the dorsal side of the body. In mature larva the bands are disappeared and change its colour to brownish grey with white spots on the body. The full grown caterpillar is measured about 45-50 mm in length. Adult moths are greyish to light brown in colour and the females are much larger than the male with a bulging abdomen. The caterpillar passes through 5-6 larval instars. Pupation generally takes place in soil around the collar region of tea bushes and shade trees. In old tea bushes, it prefers the cracks and crevices of the frame to pupate.

Hyposidra infixaria

Early instars caterpillars are brown to black in colour with transverse six white bands on the dorsal side of the body. The colour of mature larva is brownish grey with two transverse lines at the thoracic region of the body. The adults are pale or greyish in colour. The distinguishing character of this species is that the male moths bear a tuft of hairs at the abdominal tip.

Ectropis sp

The early instar caterpillars are brownish in colour with black stripes on the dorsal side of the body.The full grown caterpillar is about 30- 35 mm in length and look like a dried up twig of tea bush.

The adults are pale yellow in colour with black wavy lines. The males are smaller than the female with bipectinate antennae.

Ascotis sp

The early instar caterpillar is light green in colour. A matured caterpillar changes its colour to greenish grey with a pair of tubercle located dorsally on the abdominal region of the body.

The adults are greyish green in colour with brown spots on each wing.

Cleora sp.

The wing base of moth is white with creamy to black or even greyish black. Unlike ***Ascotis*** sp, the forewings are somewhat rounded. Fully grown caterpillars are robust with orange head and thoracic legs.

There is a pair of reddish brown tubercles dorsally and a smaller pair dorso-laterally in anterior abdominal segment of full grown caterpillar with a narrow transverse greyish bar anterior to them.

DAMAGE SYMPTOMS

The damage symptoms of different species are more or less similar. The newly hatched caterpillars make small holes along the margins of young leaves and then eat from the margins. Young ones are very often seen on growing shoots and buds scrapping on the surface also. The grown up caterpillars prefer semimature to mature leaves. From late 3rd instars, they become voracious feeder. In severity, bushes are completely stripped of the foliage. Even bark of small branches is also eaten away in absence of foliage.

RED SLUG CATERPILLARS

Leaf Infested by Red Slug Caterpillar

1. Time of Occurrence:

a. Caterpillar: Mar – Apr/May – Jun/Jul – Aug/Sep - Oct
b. Moth: Feb – Mar/May/Jul/Aug - Oct

2. Eggs and Life Cycle:

a. Oval, pale yellow, under surface of leaves & stems.
b. Hatch in 8 – 12 days in May – June.
c. Larval & Pupal stages: 4 – 5 and 2 -3 weeks respectively.
d. Life cycle: 8 – 10 weeks in summer months.

3. Damage Symptoms:

Firstly feeding on epidermal tissues and skeletonizing the lower mature leaves of the bush.

Thereafter migrating to the top hamper and feeding on leaf margins.

Leaf defoliation and bark damage in severe infestation.

BUNCH CATERPILLARS

1. Time of Occurrence:

a. Caterpillar: March – Apr/May/ - Jun/Jul – Aug/Oct - Nov
b. Pupa: April – November
c. Moth: Mar –Apr/May – Jun/Jul – Aug/Oct – Nov.

2. Eggs and Life Cycle:

a. Oval, yellow on under surface in rows.
b. Hatch in 10 – 11 days in summer
c. 8 – 9 weeks in march – June.

3. Damage Symptoms:

a. Young ones feed on epidermal tissues and margins.
b. Later Instars caterpillars devour entire leaves, remains in cluster during the day and highly active after sunset.

CONTROL MEASURES

CULTURAL CONTROL

1. Marking out of infested sections.
2. Collection and destruction of chrysalides:

   iii. Looper: Dec – Feb : Soil

iv. Red Slug: Nov – Dec: Dry leaves/ pruning litter etc.

v. Bunch: Nov – Jan: Dried Leaves

3. Trapping of moths by hand collection or light trapping:

vi. Looper: Feb – Mar and Aug - Oct

vii. Red Slug: Nov – Jan and Feb - Apr

viii. Bunch: Oct – Nov and Mar – Apr.

BIO PESTCIDES

Bacillus thuringiensis (Hatt): Dose: 800 g./ drum.

Nuclear Polyhedrosis Virus (NPV): 200 ml./ drum.

PLANT EXTRACT

Karanja extract: Dose: 200 ml./ drum

PESTICIDES

List of pesticides, which are available in the Indian market is furnished below as a ready reckoner for the tea planters. However, it is advised that only Tea Board approved pesticides should be used in the tea plantations.

MITES

1. Red spider mites *(Oligonychus coffeae)*
2. Scarlet mite (*Brevipalpus phoenicis*)
3. Pink mite (*Acaphylla theae*)
4. Purple mite (*Calacarus carinatus*)

Red Spider Mite (*Oligonychus* coffeae)

1. Time of occurrence - almost throughout the year

2. Life cycle

a. Eggs spherical and red, laid singly along mid rib & vein
b. Hatch in 4-6 days
c. Life cycle is completed in 10 – 14 days during summer.

3. Site of attack & Damage symptoms:

a. Normally attacks the upper surface of mature tea leaves. In severe cases it spreads to the undersurface and to the young leaves also.
b. Reddish spots develop on the sucking sites which later on unite to form large brown patches. Occurs mostly on the leaves upper surface.
c. The maintenance foliage turns ruddy bronze, making the infested fields distinct even from a distance.
d. In severe attack leaves may dry up and fall off.

Scarlet Mite (*Brevipalpus phoenicis*)

1. **Time of occurrence**: February – November
2. **Eggs & Life cycle**:
   a. Eggs are elliptical and deep orange to purple, slim, oval.
   b. Hatch in 11 – 13 days in March, 8 – 9 days in April and 6 – 7 days in July.
   c. Life cycle in 24 – 28 days in March, 21 – 25 days in April under laboratory conditions.
3. **Site of Attack & Damage Symptoms**:
   d. It attacks under surface of leaves at mid rib and veins.
   e. Yellowing of leaves and brownish discolouration of the mid rib mostly on the under surface of tea leaves.
   f. Sometimes splitting of stem bark may occur.
   g. Mainly young leaves are attacked.

Pink Mite (*Acaphylla theae*)

1. **Time of occurrence**: February – November
2. **Eggs & Life cycle**:
3. Eggs are colourless or light orange, carrot shape globular with shining surface, and laid singly on the surface
4. Incubation: 6 – 7 days in Dec. – Jan., 3 – 4 days in March, about 2 days in July – Aug.
5. Life Cycle: About 8 days in March, 6 days in July – Aug. and 13 days in Dec. – Jan. under laboratory conditions.

Site of Attack & Damage Symptoms:

Attacks the under surface of tea leaves.

Tender shoots are affected, damage noticeable in advance stage.

Leaf veins and edges turn pink, affected leaves become pale, dry and leathery.

Purple Mite (*Calacarus carinatus*)

Time of occurrence: February – November

Eggs & Life cycle:

a. Transparent, sub-hemi spherical, colourless and adult dark purple with white strips, 6 – 8 days in Jan., 4 – 5 days in March, 3 – 4 days in July – August.
b. Nymphal Stage: 5 – 6 days in Jan., 4 – 5 days in March - April, 3 – 4 days in July – August.
c. Life Cycle: About 7 days in July – August, 15 days in Dec. – Jan.

Site of attack & Damage symptoms:

a. Attack mostly under surface of mature tea leaves.
b. Leaves turning purplish bronze and attaining smoky grey, dry up and fall off.
c. Mite skins looks like dusting of white particles.

CONTROL OF MITES

CULTURAL CONTROL

1. Total Control of Weeds.
2. It is necessary to prevent dust, especially in the roadside blocks, by raising boundary hedges. A row of tea bushes can be left unplucked after pruning or hedge plants like *Adhatoda vasica* or *Duranta plumeiri* or *Lantana camera* can be grown.
3. Deployment of workers should be from uninfested sections to the infested sections. Entry of cattle should also be prevented.
4. Irrigation lowers the temperature thereby reduces the chance of mite infestation.
5. Less shade prone to red spider attack, since red spider mites are thermo and phototropic.
6. Planting material: China hybrids are more susceptible than Assam hybrids.
7. Prolonged drought invites red spider mite.

BIO PESTICIDES

1. *Verticillium lecanii :* Liquid: 200 ml./ drum
   Powder: 800 g./ drum
2. *Paecilomyces fumosoroseus* : Liquid:- 200 ml./ drum
   Powder:- 600 g./ drum

TEA DISEASES

All the parts of a tea plant are infected by fungus or algae. In tea, bacterial or viral disease is not a big problem. Majority of the diseases in tea is caused by fungus. It may be mentioned that the causal organisms of diseases are also pests.

PRIMARY ROOT DISEASES

Name of the disease	Causal Organism
1. Brown root Rot	*Fomes lamaoensis*
2. Charcoal Stump Rot	*Ustulina zonata*
3. Red Root Rot	*Poria hypobrunnea*

4. Black Root Rot	*Rosellinia arcuata*
5. Tarry root rot	*Hypoxylon asarcodes*
6. Purple root rot	*Helicobasidium compactum*

PRIMARY STEM DISEASES

1. Nectria	*Nectria cinnabarina*
2. Poria	*Poria hypobrunnea*

PRIMARY LEAF DISEASES

1. Black Rot	*Coticium invisum & Corticium theae*
2. Blister Blight	*Exobasidium vexans*

SECONDARY DISEASES

Secondary root diseases have pre-disposing factors for causing of the disease viz water logging, heavy drought, lightening, potash deficiency etc.

SECONDARY ROOT DISEASES

Name of the disease	Causal Organism
1. Violet root rot	*Sphaerostilbe repens*
2. Thorny blight	*Aglaospora sp.*

SECONDARY STEM DISEASES

1. Thorny blight	*Aglaospora*
2. Jew's ear fungus	*Auricularia auricular.*

SECONDARY LEAF DISEASES

1. Brown blight	*Colletotrichum camelliae*
2. Grey blight	*Pestalozzia theae.*

DESCRIPTION OF A FEW PRIMARY ROOT DISEASES

BROWN ROOT ROT (*Fomes lamaoensis*)

Roots of affected bushes are encrusted with soil , sand and pebbles, which are held by brown coloured mycelium. The crust is difficult to remove by washing and rubbing.

Soft wood inside the brown rings decays giving honey comb like appearance.

External Appearance of the Tea Bush:

Individual bush or rarely a small group of not more than nine bushes die, leaves green to reddish in colour, withered leaves remain attached.

Symptoms at the Collar Region:

Collar without encrustation but with mycelial growth.

Brown soft mycelium extending up the main stem to several inches.

Symptoms on the Root Surface:

Roots with encrustation of soil, sand, stone particles etc. The crust is difficult to remove by washing or rubbing.

Mycelium cream to brown, general colour of root mottled brown and black.

Symptom Beneath the Bark:

Films of brown mycelium on wood surface.

Symptom on the Wood Surface:

Reticulations of hard brown lines, cream coloured, soft wood between them, honey comb structure; Colour of wood light yellow.

RED ROOT ROT (*Poria hypolateritia*)

Roots are encrusted with soil , sand and stone particles held by white mycelium. Mycelium exhibits a mottled colour combination of white , red and black. Initially the encrustation is yellow but in course of time it changes to dull grey.

Roots of the affected bushes are encrusted with soil, sand and stone particles held by white & chocolate red to black cord or sheets of mycelium. The crust can be easily removed by rubbing or washing.

External Appearance of the Tea Bush :

Individual bush or rarely a small group of not more than nine bushes die, leaves green to reddish in colour, withered leaves remain attached.

Symptoms on the Root Surface :

Roots with encrustation of soil, sand, stone particles etc. Mycelium white to chocolate red or black cords or sheets. General colour of root mottled with red, black and white.

Symptoms Beneath the Bark :

Thin white films of mycelium on wood surface.

BLACK ROOT ROT (*Rosellinia arcuata*)

Mycelium is like stars, white or black. Many black dots and dashes are seen underneath the bark. Fruit bodies are black in colour and spherical in shape.

External Appearance of the Tea Bush:

Individual bush or rarely a small group of not more than 9 bushes die, leaves green to reddish in colour, withered leaves remain attached.

Symptom at the Collar Region :

Collar without encrustation but with mycelial growth. Woolly, greyish stocking of mycelium sometimes extending to the base of the branches.

Symptom on the Root Surface :

Roots with cords or strands of mycelium on the surface. Cords cobwebby, black or greyish black, loose or adherent.

Symptom Beneath the Bark :

Star shaped , small, white to black strands of mycelium accompanied by irregular dots and dashes.

Symptoms on the Wood Surface:

Wood with black dots and dashes.

Special Character

1. Star like mycelium, white to black, accompanied by numerous small dots and dashes underneath the bark.
2. Black, spherical fructifications.

TARRY ROOT ROT (*Hypoxylon asarcodes*)

A black encrustation at the collar region like a smooth and hard black crust.. The mycelium gives a beautiful graphical representation.

External Appearance of the Affected Bush :

Individual bush or rarely a small group of not more than nine bushes die, leaves green to reddish in colour, withered leaves remain attached.

Symptoms at the Collar Region:

Collar with encrustation (sand, stones, pebbles).Encrustation black, effused, hard surface of crust smooth and like dried black point.

Symptoms on the wood :

Wood with dark bands and small reticulations of light brown thin lines.

Special Characters

1. The stems from the ground level upward to several inches develop a black encrustation. Graphs like *Aglaospora*.
2. Collar carries a smooth hard crust.

PURPLE ROOT ROT (*Helicobasidium compactum*)

Thick velvety mycelium is found on the colour region. Affected wood turns pink in colour.

External Appearance of the Affected Bush:

Individual bush or rarely a small group of not more than nine bushes die, leaves green to reddish in colour, withered leaves remain attached.

Symptoms at the Collar Region:

Thick pad of velvety, purplish brown mycelium surrounding the collar.

Symptoms at the Root Zone:

Cords of mycelium purplish brown, cords branched, giving a general purplish brown colour to the root.

Symptoms on the Wood Surface:

Woods without bands, reticulations or mycelium. Colour pink to pinkish brown.

Special Character

A thick velvety mycelial pad just above the soil level often surrounding whole stem to several inches. Colour of the affected wood is pinkish.

CHARCOAL STUMP ROT (*Ustulina zonata*)

When the bark is peeled off, white , silky mycelium is visible. Charcoal like black brittle encrustation is found on the collar . Fruit bodies develops on the collar region.

External Appearance of the Affected Bush:

Mature bush part dead and remainder apparently normal and healthy. Dead part may be completely rotten.

Symptoms at the Collar Region:

Encrustation at first white, then greenish grey, finally black. Surface undulated, brittle, charcoal like when broken.

Symptoms at the Root Zone:

Roots without cords or strands of mycelium on the surface. White or black cushion or lumps on the root.

Symptoms Underneath the Bark:

Film of dull white, silky, fan like mycelium.

DESCRIPTION OF A FEW SECONDARY ROOT DISEASES

VIOLET ROOT ROT (*Sphaerostilbe repens*)

This is a secondary disease and develops in the area which is waterlogged. The roots become inky black or light violet coloured. The smell of the dug root is vinegary. Leaves of the affected bushes drop off in green condition only. Fruit bodies develops at the collar region in clusters and resembles tiny flower buds.

SHOW STRING FUNGUS (*Armillaria mellea*)

A split develops at the collar region upto 6 cm from the ground level. This furrows of the split is covered by white mycelium.

Roots posses brownish black rhizomorphs which looks like shoe strings.

THORNY BLIGHT (*Aglaospora syn Tunstallia aculeata*)

After the bark is peeled off, fern like mycelium is seen. It is called thorny blight because the fruit bodies , which develop on the bark has thorn like projections.

DESCRIPTION OF A FEW STEM DISEASES

THORNY STEM BLIGHT (*Aglaospora syn Tunstallia aculeata*)

White strands of mycelium is seen on the surface underneath the bark. Black, thorn like projections are seen on the bark.

BRANCH CANKER (*Poria hypobrunnea*)

Thin irregular brown coloured lines are seen on the wood. Infected wood becomes yellow coloured . It becomes soft and ultimately decays. Fruit bodies are yellow coloured, which turns dark grey in course of time. Fruit bodies are found at the base of the big branches.

STEM RED RUST (*Cephaleuros parasiticus*)

Red rust lesions are circular in shape. The fructifications are brick red or orange coloured. Sporangia are located at the apex takes the shape of a knob in which numerous spores are present.

A FEW LEAF DISEASES

BLACK ROT (*Corticium invisum / Corticium theae*)

The apex and margins turns black and the leaf become sodden in the rainy season. The dry leaves develops a mottling of brown and grey colours. Two or more leaves remain attached with the help of a gummy exudation.

LEAF RED RUST (*Cephaleuros mycoides*)

Small brownish warts develop on the leaf surface which are circular in shape.

BLISTER BLIGHT (*Exobasidium vexans*)

Swallow depression is found on the upper surface of the leaves , which appears like convex warts on the under surface. Greyish white spores are found inside these warts.

BROWN BLIGHT (*Colletotrichum camelliea*)

Generally old leaves are affected by this disease. On young leaves also yellowish green spots develop. Circular irregular spots develop on the leaves with grey and red mycelium.

GREY BLIGHT

The patches on the leaves are light brown (sometimes dark brown). The shape of the patch is circular or oval. The concentric rings of mycelium are distinctly visible. On the dorsal surface of the leaves, fruit bodies (pustules) are located in concentric rings. The size of the fruit bodies are slightly bigger than the fruit bodies of Brown blight.

CONTROL OF PRIMARY ROOT DISEASES

The control measure varies according to the condition and severity of infection.

Condition I : When only a single individual bush is infected , the infected bush should be immediately removed. Treat the soil with fungicides.

Condition II : When few bushes are infected in a small patch, remove all the affected bushes and also a row of healthy plants surrounding the patch. Treat the soil with fungicides.

Condition III : When a large area is affected, isolate the area by digging a 90 cm deep trench circling the affected area. The dug soil should be spread inside the ring. Uproot all the bushes and treat the soil with fungicides.

CONTROL OF FUNGAL OR ALGAL DISEASES IN GENERAL:

Spray fungicides on the affected bushes as per recommendation given on the fungicide list.

A FEW PERMISSIBLE ACARICIDES,INSECTICIDES & FUNGICIDES

FORMULATION	ONE EXAMPLE	DOSE PER HECTARE
ACARICIDES		
Cyflumetofen 20 %SC	FOSTER	750 ml
Ethion 50 %EC	ETHION	1000 ml
Fenazaquin 10 %EC	MAGISTER	1000 ml
Fenpyroximate 5 %EC/SC	PYROMITE	400 ml
Hexythiazox 5.45% EC	MAIDEN	400 ml
Propargite 57 %EC	SIMBA	500 ml
Spiromesifen 22.9% SC	OBERON	300 ml
Etoxazole 10 %SC	BORNEO	250 ml
Flufenzine 20 %SC	FLUMITE	450 ml
Propargite 42% +Hexythiazox 2%EC	CEASEMITE	1000 ml
Fenazaquin 10% + Bifenthrin 4% EC	MAGNITUDE	600 - 1000 ml

FORMULATION	ONE EXAMPLE	DOSE PER HECTARE
INSECTICIDES		
Bifenthrin 8 %SC	MARKAR	750 ml
Clothianidin 50 %WDG	DANTOTSU	80 gms
Deltamethrin 2.8 %EC	DECIS	500 ml
Dimethoate 30 % EC	ROGOR	750 ml
Emamectin Benzoate 5 %SG	POCKET	200 gms
Emamectin Benzoate 3% + Thiamethoxam 12% WG	ENCOUNTER	200 - 300 gms
Fenpropathrin 30% EC	MEOTHRIN	250 ml
Flubendiamide 20 %WG	TAKUMI	80 – 140 gms
Flubendiamide 19.92% w/w + Thiacloprid 19.92% w/w SC	BELTEXPERT	250 ml
Quinalphos 25 %EC	EKALUX	1000 ml
Thiacloprid 21.7 %SC	ALANTO	500 ml
Thiamethoxam 25 %WG	ACTARA	100-125 gms
Thiamethoxam 12.6% + L-Cyhalothrin 9.5%	ALIKA	150 ml

FORMULATION	ONE EXAMPLE	DOSE PER HECTARE
FUNGICIDES		
Carbendazim 12% + Mancozeb 63% WP	COMPANION	400 gms
Hexaconazole 4% + Zineb 68% WP	AVTAR	500 - 600 ml
Hexaconazole 5% EC	CONTAF	200 ml
Propiconazole 25% EC	TILT	125 ml
Copper Oxychloride 50% WP	BLITOX	420 gms
Tetraconazole3.8%w/w(4%w/v)	DOMARK	200 -400 ml
Trifloxystrobin25%+Tebuconazole 50% WG	NATIVO	100 gms

Exercise No. 11

Weeds & Weed Control

The term weed is applied to " all those plants for which man has found no use and which in their evolution have developed such power of aggression, persistence or reproduction as to make them a menace to the best possible development of crop. Weed can also be defined as certain undesirable species of plants which are unwanted, un useful often prolific and persistently interfere with agricultural operation , increase labour , add to cost and reduce crop yield.

HARMFUL EFFECT OF WEEDS

1. In young tea weeds can remove as high as 270 kg of Nitrogen per hectare.
2. Few species of weeds serve as alternate hosts of pests .
3. In Sri Lanka 9 % crop loss have been reported in tea industry.
4. In Indian tea, 6 to 12 % crop loss has ben reported.
5. Weeds competes with tea for nutrients, moisture and light.
6. Weed infestation can restrict branching and frame development in young tea
7. Weed coverage on plucking surface disrupts photosynthesis and productivity.
8. Weed hindrance on the path of the pluckers reduce pluckers' productivity.
9. Reduce water flow in the drains.

TYPES OF WEEDS

Annuals: 1. Complete life cycle within one season of two to four months.

2. Produce large quantity of seeds.

3. Growing period coincides with flushing period of tea.

eg-*Ageratum conyzoides*

Biennials: 1. Complete life cycle in two years.
2. Seed settling and flowering starts from second year.
eg-Cassia sophera

Perennial: 1.They are persistent type.
2. Live more than two years.
eg- Axonopus compressus

For Chemical Control of Weeds, they are Categorised in Two Groups:

MONOCOTS (mainly grasses)

DICOTS (broad leaved weeds)

CRITICAL PERIOD OF WEED COMPETITION

The critical period of weed competition is tea is April to September. Utmost care is needed to be taken to control weeds during this period so that the productivity of tea is not affected.

CONTROL METHODS

1. Cultural method
2. Manual /Mechanical method
3. Mechanical method
4. Biological method
5. Chemical method

CULTURAL CONTROL

MULCHING

Mulch is a protective covering of material maintained on soil surface. Mulching has smothering effect on weed control by excluding light from the photosynthetic portions of a plant and thus inhibiting the top growth

RAISING COVER CROPS

Mainly in young tea after land preparation green cropping or cover cropping is done.

Cover crop is generally a leguminus plant having "nif" genes, so it can fix nitrogen in soil

It privides organic matter to the soil. Beside increasing soil fertility , it also reduces weed growth.

MANUAL / MECHANICAL CONTROL

HAND WEEDING

It is done by physical removal or pulling out of weeds by hand .

SICKLING

It is also done by hand with the help of sickle to remove the top growth of weeds to prevent seed production and to starve the underground parts

TILLAGE

Tillage removes weeds from the soil resulting in their death. It may weaken plants through injury of root and stem pruning, reducing their competitiveness or regenerative capacity.

HOEING

Hoe has been the most appropriate and widely used weeding tool for centuries. Zt is however, still a very useful implement to obtain results effectively and cheaply.

CHEEL HOEING

Cheel hoeing or cheeling using a half moon hoe is mostly followed in young tea is not only expensive but often leads to greater infestation of weeds. Primarily it is due to exposure of the top soil layer and the weed seed present therein. Exposure of soil also leads to soil erosion.

BIOLOGICAL CONTROL

Biological control is done by using living organism's viz., insects, disease organisms, herbivorous fish, snails or even competitive plants.

In biological control method, it is not possible to eradicate weeds but weed population can be reduced. This method is not useful to control all types of weeds. This method is still in the experimental stage.

CHEMICAL CONTROL

Chemical control is done by using chemical herbicides.

PRECAUTIONS TO BE TAKEN DURING PESTICIDE SPRAY

- Use apparels and masks.
- Storage herbicide in isolated place away from other chemicals.
- Mixing done in open place and in plastic drums.
- Keep grazing animals away from the spraying zone.
- Washing of sprayers done on non cultivated land area.

CHARACTERISTIC FEATURES OF A FEW MAJOR WEEDS

Abutilon indicum L. - Family : Malvaceae

It is a perennial herb, goes upto a height of about 3 metres.

Stem covered with soft velvety tomentose.

Leaves ovate or orbicular, The length of the leaf on an average is 7 cm and width about 5 cm.

Shape of the leaf is cordate, apex is acute or acuminate, entire.

Stipules small ; colour of the flowers is orange or golden yellow about 2.5 cm diameter.

Flowering & Fruiting period : Rainy season and winter

Acaphyla indica L. – Family : Euphorbiaceae

Annual herb about 25-75cm tall;

Leaves are about 3 to 6 cm long and 2 to 3.5 cm wide. Shape of the leaf is generally ovate., apex acuminate, margin crenate serrate;

Flowers are green coloured. Inflorescence is a spike. Male flowers are clustered or few towards the top; females flowers are found on the at the lower side; Flowers are surrounded by bracts

Flowering & Fruiting period : Summer and rainy season

Achyranthes aspera L. – Family : Amaranthaceae

This weed can be annual or perennial. Height is about 60-90cm.;

Leaf: Shape of the leaf is ovate or broadly elliptic; Length goes upto 10cm and the width is about 5 cm., Apex is acuminate or sub-acute.; Pubescense is tomentose, velvety or glabrate;

Flowers: Generally pale green in colour often tinged with purple. Inflorescence is a panicle or spike. Bracts are ovate in shape, apex is acuminate, persistent , Bracteoles concave with a spine, utricle oblong, cylindrical.

Flowering & Fruiting period : Throughout the year

Ageratum conyzoides L. – Family : Asteraceae

An annual herb. The height goes upto 80 cm.

Leaves opposite at the bottom and alternate on the upper hamper, length is about 8 cm . Width is about 2 to 5cm, Shape of the leaf is ovate, apex is acute, hairy; margins crenate,

Flowers white, blue or light purple coloured; arranged in discoid heads,

Flowering & Fruiting period : October to April

Amaranthus sessilis (L.) DC – Family : Amaranthaceae

It is a perennial herb, height 30- 80cm.

Leaves opposite, oblong or linear oblong, length is 2 to 5cm and width about1.5cm, apex is acute or acuminate, margin is entire;

Flowers are whitish or pinkish in colour, arranged in axillary 1-4 sessile heads;

Flowering & Fruiting period : July to March

Amaranthus spinosus L. – Family : Amaranthaceae

It is an annual herb, about 60-90 cm tall. Sharp axillary spines present;

Phyllotaxy of the leaves is alternate, shape ovate to lanceolate, length 4 to 9 cm , width 2.5 to 5.0 cm, margin is entire, apex is acuminate and base is cuneate.

Flower cluster dense, lower axillary and upper in axillary and terminal spikes;

Colour of the seeds is black or brownish black

Flowering & Fruiting period : Rainy season and post rainy season.

Agremone Mexicana L. – Family : Papavaraceae

It is an annual herb. Emits yellow latex, height 60-90 cm.

Leaves sessile, half-amplexicaul, length of the leaf is about 10 to 15 cm. , lobes dentate with yellow prickles. Flowers are golden yellow in colour, sessile, solitary; capsule oblong, elongated, usually prickly.

Flowering & Fruiting period : October to April.

Artemisia vulgaris L. – Family : Asteraceae

It is small perennial shrub with aromatic pubescent hair. Height is about 1.5 metres.

Leaf is grayish green coloured. Length of the leaves is about 5-10 cm , shape is ovate with stipule like lobes at the base,

The shape of the leaf is linear or linear-lanceolate,

Flowers are brownish-yellowin colour, sessile or shortly pedicelled; has globose or sub globose heads;

Flowering & Fruiting period : August to October

Bidens biternata (Lour) Merr & Sherff – Family : Asteraceae

It is an annual herb about 0.6 m tall.

Leaves are opposite, pinnately compound 3-7 lobed, the shape of the leaflets ovate or lanceolate, serrate;

Heads on long stout peduncles, corymbosely panicled;

Ray florets are sterile and white in colour.

Disk florets are yellow in colour and are bisexual in nature.

Flowering & Fruiting period :May to November.

Borreria articularis (L.f.) F.N.Will – Family Rubiaceae

A branched herb having a height of 20-60 cm.

Branches quadrangular, hairy;

Leaves 1.5 -4.0 cm long and 1 to 1.5 cm wide, sessile, shape of the leaf is elliptic-oblong;

Apex is obtuse or acute . coriaceous;

Stipules present;

Flowers whitish blue in colour, generally 4-6flowers arranged in a whorl .

Flowering & Fruiting period : July to November.

Croton bonplandianum Bail. – Family : Euphorbiaceae

It is a woody herb with elongated lenticels; body is covered with dense hair.

Leaves are generally green in colour and the shape is lanceolate. Length varies from 2.5 to 6.0 cm. It has a short petiole;

Flowers greenish white in colour arranged in racemes, males flowers are on the upper side and female flowers are located at the bottom. Large glands; are present. Capsule is oblong, obtuse, triquetrous, stellately hairy.

Flowering & Fruiting period :Summer and Winter.

Cuscuta reflexa Roxb. – Family : Convolvulaceae

It a perennial parasitic herb. Stem is pale yellowcoloured, fleshy and glabrous;

Flowers are small, white in colour, pedicelled, solitary or in clusters of 2-4 or in small racemes;

Bracts are oblong or obtuse;

Capsule is globose.

Flowering & Fruiting period :January to April.

Datura metel L. – Family : Solanaceae

This plant may be biennial or perennial small shrub about 0.8-1 metre tall.

Leaves are ovate, ovate-lanceolate, apex acute, margin dentate;

Flowers generally white in colour , solitary, erect terminal, pedicelled;

Capsule globose and prickly.

Flowering & Fruiting period : October to March

Drymaria cordata Willd. – Family : Caryophyllaceae

It is a glabrous and dichotomously branched herb.

Leaves are ovate-cordata or orbicular with stipules.

Flowers are white in colour arranged in axillary or terminal long peduncle cymes;

Capsule ovoid, trigonous3-valved to the base.

Flowering & Fruiting period : October to February

Eupatorium odoratum L. - Family : Asteraceae

It is a profusely branching perennial herb.

Leaves are opposite, petioled, aromatic, shape of the leaf is ovate-deltoid, length of the leaf is about 6 to 9cm and 2.5 to 5cm. wide, apex acuminate, sparingly serrated;

Flowers are white in colour ; Inflorescence is corymbose.

Flowering & Fruiting period : November to March.

Euphorbia hirta L. – Family : Euphorbiaceae

It is a small prostrate annual herb which emits latex. There is a huge variation in the length of the herb ranging from 15 cm to 45cm . On the body long yellowish hair is present.

The phyllotaxy of the leaves is opposite, length is 1.5 to 3.5cm , shape is elliptical oblong, obovate to lanceolate, shortly stalked, apex acute, margin dentate o, upper leaves are dark green in colour while lower leaves are pale green.

Flowers are arranged in axillary or terminal raceme; Capsules hairy with reddish-brown seeds, trigonous, minutely rugose.

Flowering & Fruiting period : Throughout the year

Galinsoga parviflora Cav. – Family : Asteraceae

It is an annual herb , Height is about15 to 30 cm.

Leaves opposite, entire or toothed;

Heads small, peduncled, sub-terminal and axillary, achenes angled or compressed.

Flowering & Fruiting period : Winter

Gnaphalium indicum L. - Family : Asteraceae

It is an annual herb having white wooly hair on the body.

Leaves are obovate to ovate-lanceolate in shape.

Raceme heads are white in colour and sessile.

Flowering & Fruiting period : December to April

Lantana camara L. – Family : Verbenaceae

It is a shrubby weed.

Stem is quadrangular covered with re-curved prickles;

Leaves are opposite, shape is ovate, obovate or oblong, length varies from 2.5 to 9 cm and width varies from 1 to 4 cm wide , apex acute , margin crenate, rounded or cordate at the base;

The colour of the flower is orange red, scarlet or varying from white to yellow colour, . Inflorescence is corymbose.

Flowering & Fruiting period : Throughout the year

Leucas aspera (Willd.) Spreng. – Family : Lamiaceae

A hairy aromatic herb , 30- 50cm tall.

Stem is quadrangular;

Leaves subsessile or short petioled, linear, narrowly oblong or oblong-lanceolate, entire or distantly crenate, narrowed at the base;

Flowers are white coloured, sessile in terminal and axillary, very dense whorls;

Flowering & Fruiting period :Throughout the year

Limnophila heterophylla Benth. – Family : Scrophulariaceae

This is an aquatic weed, 30-40cm in height.

Leaves are pinnately compound , opposite, small, sessile. Shape of the leaf is oblong. Lower leaves are whorled.

Flowers are pinkish or bluish, solitary, sessile or sub-sessile;

Fruits sub-orbicular or hemispheric.

Flowering & Fruiting period : October to March.

Lindenbergia indica (L.) Kurz. - Scrophulariaceae

It is an aromatic annual herb.

Shape of the leaf is ovate- obtuse, Apex crenate- serrate. Leaf is glandular villous on both surfaces, Petiole is long.

Flowers are yellow with red or purple tinge.

Pedi-celled in axillaries or terminal spicate racemes;

Capsule is ovoid.

Flowering & Fruiting period :December to April

Melastoma malabathricum L. – Family : Melatomaceae

It is a bushy shrub, generally grows near the water bodies / ditches.

Leaves are opposite, elliptic or lanceolate or oblong- lanceolate in shape. The length of the leaf is about 7 to 12 cms and width abou 3 to 5 cm. Body is covered with hair, petiole is strigose. Flower is purple in colour and grows in clusters.

Bracts encloses the buds. Fruit is short, ovoid and truncate.

Flowering & Fruiting period : Throughout the year

Mikania cordata (Burm. F.) L. Robinson - Asteraceae

It is a climber which covers the plucking surface of tea plants inflicting considerable crop loss.

Leaves have long petiole. Phyllotaxy opposite, Shape cordate or ovate-cordate, apex accute or acuminate.

Flowers are white in colour , small heads surrounded by an involucres bracts on a common disc.

Flowering & Fruiting period : November to January

Mimosa pudica L. – Family : Mimosaceae

A small herb with thorns. This plant is also called Touch Me Not. It is sensitive and the leaves closes if touched.

Leaves pinnately compound, Pinnae- 4 , distantly arranged, 2.5-10.0 cm long.

Petiole bristly, stipules linear- lanceolate with bristly edges . Leaflets 12-20 pairs obliquely narrow-oblong, under surface bristly.

Flowers are pink in colour situated on slender axillaries peduncles, pods jointed and prickly.

Flowering & Fruiting period : August to December.

Oldenlandia corymbosa (L.) Lamk – Family : Rubiaceae

It is an annual herb, 20-40 cm tall.

Stem quadrangular, slightly scabrid on angles, sparsely hairy.

Phyllotaxy of the leaves is opposite. Shape is linear or narrowly elliptic or elliptic- laceolate, apex acute, slightly pubescent beneath, narrowed at the base.

Stipules have sheaths. Flowers are white, solitary or 2-4 flowered in peduncled axillary cymes. Capsule small and ovoid.

Flowering & Fruiting period : July to December

Phyllanthus niruri L. – Family : Euphorbiaceae

It is an annual herb about 60 cm tall.

Leaves are sessile or sub-sessile.Phyllotaxy imbricate.

Leaflets are small ,1 to 1.5 cm long. Shape elliptic, obovate or oblong. Apex rounded, obtuse or acute.

Flowers are yellowish , axillary and with short pedicels.

Capsule globose, smooth, obscurely lobed.

Flowering & Fruiting period : August to October

Polygonum chinense L. – Family : Polygonaceae

It is a perennial herb found near swampy areas.

Leaves are alternate, shape ovate- cordata, or lance late. Length of the leaf varies from 2.5 to 7.5 cms. Apex is acute or acuminate, with a dark mark on the upper surface towards the basal part. petiole slightly auricled.

The colour of the flower is light- pink or white, situated on axillary or terminal peduncled heads.

Fruit a trigonous nutlet.

Flowering & Fruiting period : September to March

Polygonum hydropiper L. – Family : Polygonaceae

It is an annual herb 30 to 60 cm tall.

Stem is branched, minutely glabrous, swollen at the joints.

Leaves are sub- sessile, linear to linear-lance late or oblong in shape , length is about 6 to 10cm and width about 1.5 to 2 cm, apex is acute or acuminate, gland dotted, scabrous on the margins.

Flowers are pink or pinkish white, decurved filiform, flexuous racemes.

Nut lets triquetrous, opaque, granulate.

Flowering & Fruiting period : October to March

Scoparia dulcis L. – Family : Scrophulariaceae

It is an annual herb or under shrub, 40 to 70 cm tall .

Leaves opposite or alternately whorled, glandular, shape of the leaves is elliptic or rhomboid, 3 to 4 cm long and 1.5 to 2 cm wide. Petiole is short.

The colour of the flower is white or whitish blue, axillary, solitary or in racemes.

Capsule small, ovoid or globose.

Flowering & Fruiting period : July to January

Sida acuta Burm.f. – Family : Malvaceae

It is a branched herb or undersrub, 40 to 70cm tall.

Branches are hairy.

Leaves are lanceolate to linear, rarely ovate to oblong. Stipules unequally paired at the node.

Flowers are yellow in colour and arranged on axillary pedicels; capsule strongly reticulate.

Flowering & Fruiting period : September to January

Solanum nigrum L. – Family : Solanaceae

A perennial herb or under shrub.

Shape of the leaf is ovate-oblong or ovate-lanceolate ,length often 3 to 10 cm and width 4 to 6 cm. Apex is acute or acuminate, glabrous, attenuated at the base.

Flowers are white in colour , arranged in axil, long peduncled, umbellate cymes.

Berries globose, black or red and shining when ripe.

Flowering & Fruiting period : March to November

Solanum torvum Swartz. – Family : Solanaceae

It is a perennial tomentose shrub 1.5 to 2.5cm tall, with thorns and sparsely stellate hairs.

Leaves are ovate or ovate-oblong, 7.5 to 15 long and 6 to 12 cm wide. Apex is acuminate, sparsely stellate tomentose above and densely beneath, sparsely prickly on the petiole and 1-2 prickles on the midrib beneath.

Flowers are white in colour and arranged in extra-axillary dense dichotomous cymes.

Berries are yellow in colour and globose.

Flowering & Fruiting period : September to December.

Sonchus arvensis L – Family : Asteraceae

This is an annual herb which exudes milky juice, about 30 to 90 cm. tall with creeping stoloniferous root-stock.

Stem hollow, angular, umbellately branched.

Leaves are runcinate, pinnatifid, spinous toothed with alternate phyllotaxy.

Heads are pale yellow to yellowish, white, umbellately corymbose.

Flowering & Fruiting period : October to February

Tridax procumbens L. – Family : Asteraceae

A slender herb with flower on top.

Leaves are ovate or ovate-lance late, pinnatifid, hairy.

Flower heads are creamy yellow or white yellow.

Achenes obovate or oblong, silk brown, slightly 3- angled.

Flowering & Fruiting period :Throughout the year.

Urena lobata L. – Family : Malvaceae

It is an undershrub. 1to 1.5 metres tall.

Leaves are ovate to orbicular, 7.5 to 12.5 cm. long, lobed pubescent beneath, midrib with a linear gland near the base.

Stipules lanceolate to obovate.

Flowers pink in colour with a darker center, axillary solitary or in a cluster.

Capsule globose, covered with stellate barbed bristles.

Flowering & Fruiting period : July to January

Axonopus compressus (Sw)Beauv. Poaceae

It is a perennial grass with stolons.

Leaves are linear, flat, oblong – lanceolate, apex obtuse or acute, 5 cm to 10 cm long and 0.5 cm to 1.5 cm broad., glabrous or sparingly hairy; spikelets oblong or oblong –lanceolate, sub acute or acute, sessile, erect or spreading racemes; grain oblong elliptical and white.

Flowering & Fruiting period : August to December.

Commelina benghalensis L. Commelinaceae

It is a small dichotomously branched herb.

Leaves are sessile or short petioled, elliptical – ovate in shape. 3 cm to 8 cm long and 1 to 3 cm broad., apex obtuse, rounded at the base; sheaths pubescent or villous,

Flowers are blue in color, cyme, 2-3 flowered enclosed by spathe , turbinate; capsule membranous, 5 seeded.

Flowering & Fruiting period : September to December

Cynodon dactylon (L.) Pers. Poaceae

It is a perennial creeping grass and forming matted tufts.

Leaves are lanceolate to linear lanceolate, apex acute or acuminate, 2 to 12.5 cm long and 3 mm wide, glaucous, glabrous or hairy below, spikes 4 to 5 , spikelets green or purplish , grains oblong free within lemma and palea.

Flowering & Fruiting period : December to February

Cyperus rotundus L Cyperaceae

It is a herb with rhizome , 10 to 75 cm tall.

Leaves are linear , tapering, acuminate;

Umbels are simple or compound, 2-8 rayed; bracts 3 , leaf like, unequal;

Spikelets are pale – yellow or brown coloured with a reddish tinge; glumes oblong, obtuse or apiculate; nut trigonous, roadly obovoid, greenish black

Flowering & Fruiting period : July to November

Eleusine indica (L.) Gaertn. Poaceae

This is an annual grass with culms erect or geniculate, 10 to 75 cm long , 2-3 noded.

Lleaves crowded at the base, linear, 10 cm to 30 cm long and 0.3 to 0.5 cm broad,erect glabrous or sparingly hairy; ligule minute or absent; spikes 2 to 7, digitate, 2.5 cm to 12.5 cm long.

Spikelets sessile, in two rows, ovate or ovate-oblong, green coloured, 3- 6 flowered; grains oblong, obtusely trigonous, transversely rugose.

Flowering & Fruiting period :October to February.

Imperata cylindrical (L.) P.Beauv, Poaceae

It is a perennial grass, 0.8 metre to 1.2 metres tall

Root stock stoloniferous and widely creeping; culms leafy, solid, slender, thickened, pubescent or wooly at base; nodes densely bearded with erect white hair,

Leaves narrow, linear, flat, apex acuminate; ligules ciliate, silky at the base;

Panicle 10 cm to 20 cm long; spikelets lanceolate.

Flowering & Fruiting period :March to August

Paspalum scrobiculatum L Poaceae

It is an annual grass with fibrous root system. Culms erect, decumbent at the base, 25 cm to 80 cm tall, tufted.

Leaves are linear, 12.5 to 25 cm long, apex acuminate; ligule short, membranous;

Panicle erect with 2 to 4 spikes, 4 cm to 10 cm long. Spikelets in 2 rows, imbricate , orbicular or ovoid, glabrous. Grain bi convex.

Flowering & Fruiting period : April to June

Setaria glauca (L.) Beauv. Poaceae

It is an annual grass , 3 cm to 60 cm tall, loosely tufted.

Lleaves are flat, tapering to a slender point, glabrous or sparsely hairy, margins scabrid.

Flowers are in cylindrical spike, covered with stiff brown hair.

Flowering & Fruiting period : May to November

A FEW PERMISSIBLE HERBICIDES

FORMULATION	ONE EXAMPLE	DOSE PER HECTARE
HERBICIDES		
Glyphosate 41%SL	ROUND UP	2000 – 3000 ml
Glyphosate 71%SG	MERA 71	1500 – 2000 gms
GlufosinateAmonium 13.5% SL	SWEEP POWER	3000 ml
Oxyfluorfen 23.5% EC	OXYGOLD	400 – 800 ml
Paraquat Dichloride 24%SL	GRAMOXONE	2250 ml
Oxyfluorfen2.5%+Isopropylamine salt of Glyphosate 41%w/wSC	ADAMA	2000 ml
Indaziflam1.65%w/w+Glyphosate-isopropylammonium44.63%w/wSC	ALION PLUS	2500 ml
Saflufenacil 70% WG	DUVELON	100 gms
Triasulfuron20% WG	LOGRAN	100 gms
2,4-Daminesalt 58%WSC	WEEDNASH	1400 gms
GlufosinateAmmonium13.4%+ Oxyfluorfen 4.8% w/w	FASCINATE	2400 ml

COLOUR PLATES

BOTANICALS

Ageratum Conyzoides

Artemisia Vulgaris

Equisetum Arvense

Polygonum Runcinatum

Urtica Dioecia

PICTURES OF NUTRIENT DEFICIENCY

Zinc, Magnesium, Nitrogen and Potash Deficiency

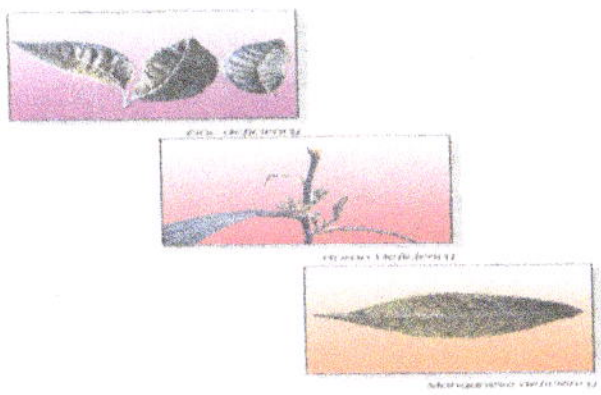

(Courtesy : Tocklai Experimental Station and T. Stanes & Co.)

Acute Magnesium Deficiency

PICTURES OF SPRAYERS

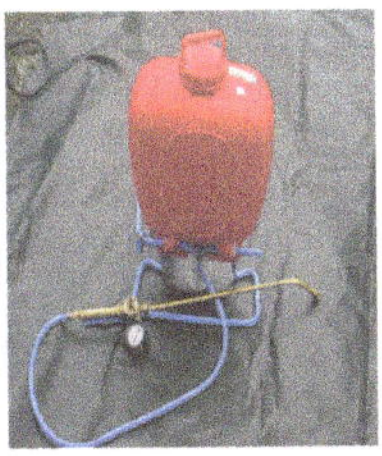

Aspee Bapak Herbicide Sprayer

Aspee Bakpak Insecticide-Foliar Sprayer

Aspee Bolo Power Sprayer

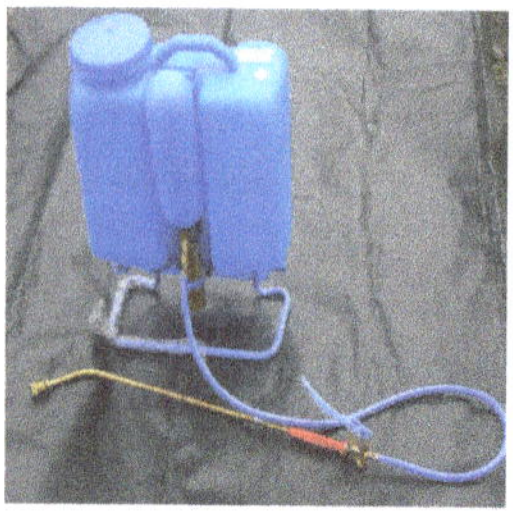

Aspee Hitech Sprayer

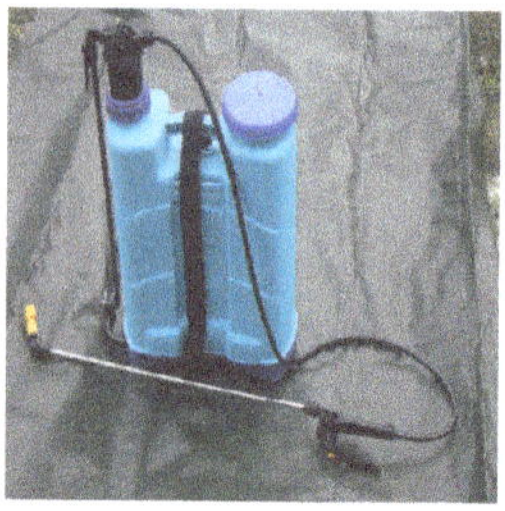

Aspee Knap Sack Sprayer

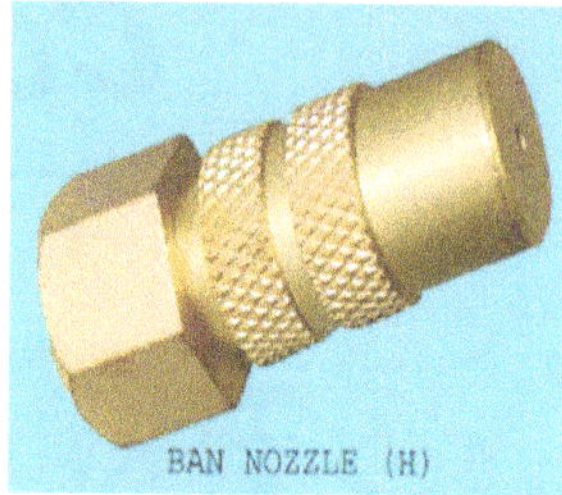

Ban Nozzle

Battery Operated Sprayer

Electrically Operated Sprayer

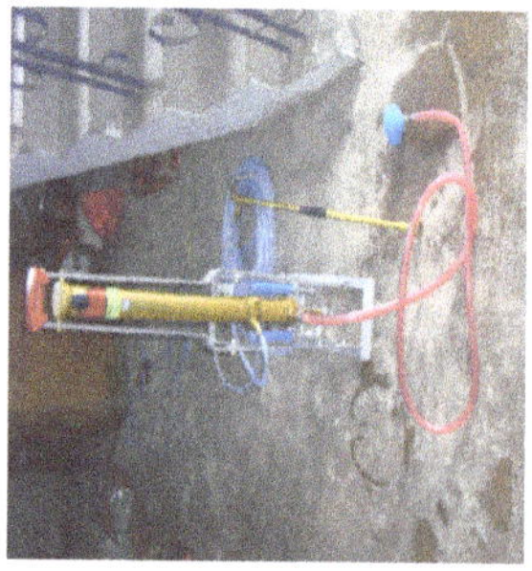

Foot Sprayer For High Tree Spray

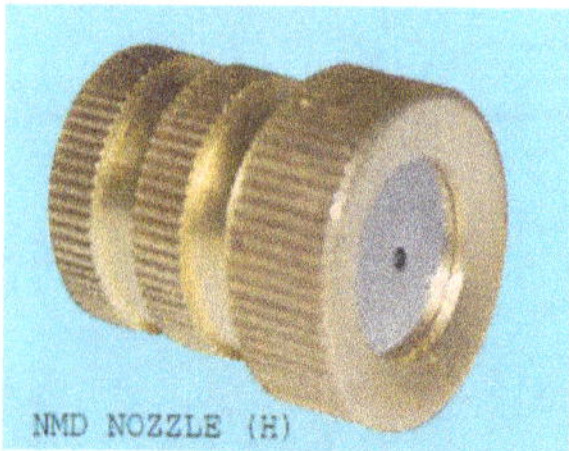

NMD Nozzle

Tractor Mounted Sprayer

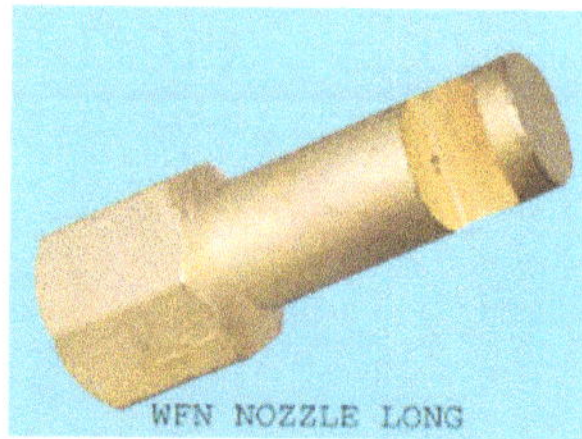

WFN Nozzle

PICTURES OF ENTOMOPATHOGENS & PHYTOPATHOGENS

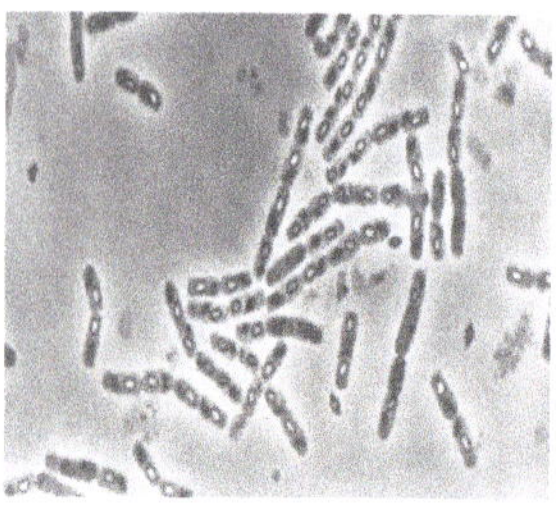

Bacillus thuringiensis

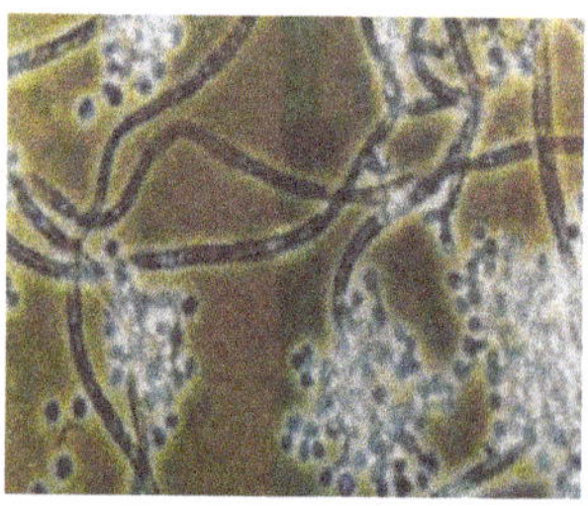

Beauveria Bassiana

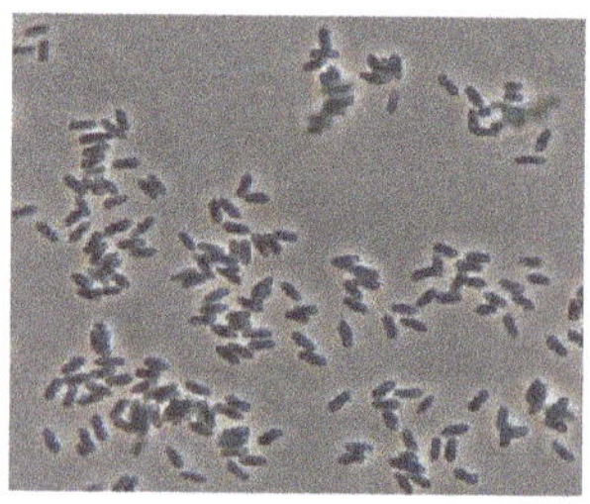

Metarhizium Anisopliae

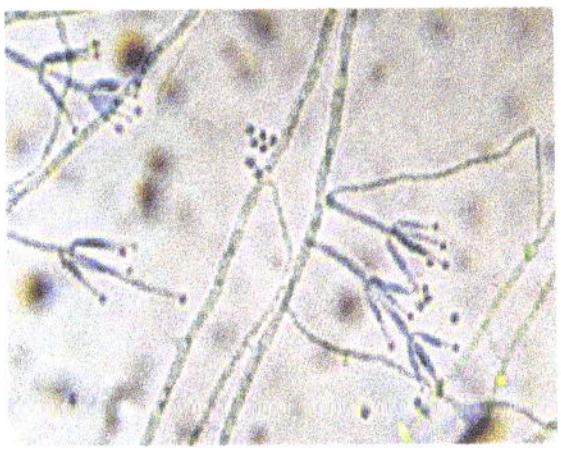

Paecilomyces Fumosoroseus

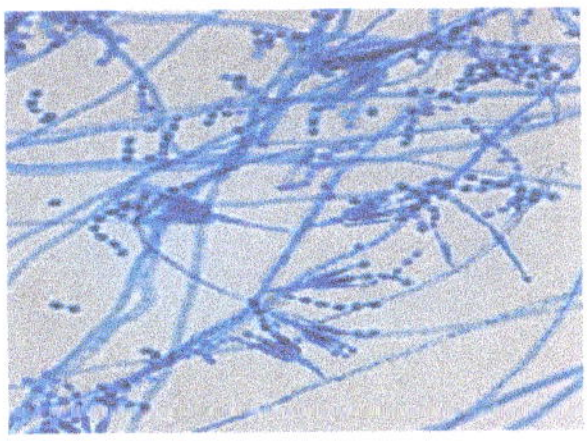

Paecilomyces Lilacinus

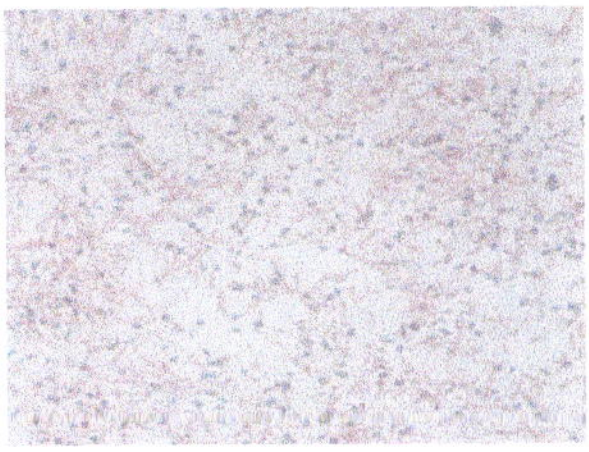

Pseudomonus Fluorescence

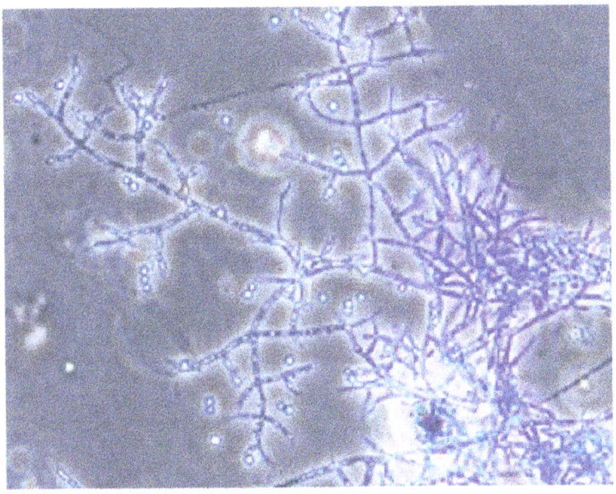

Trichoderma Viride

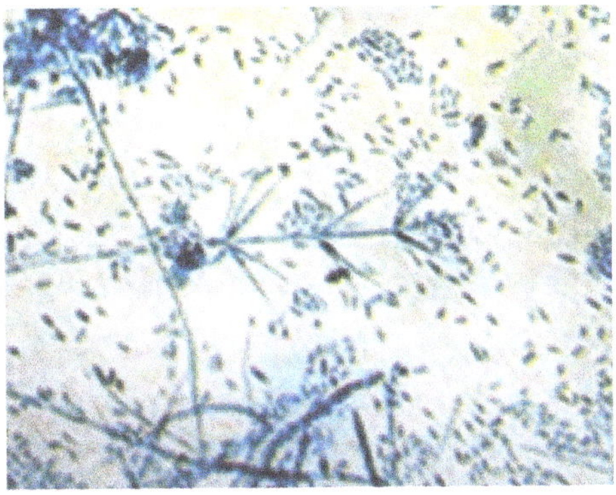

Verticillium Lecanii

PICTURES OF INSECTS

Biston suppressaria moth (Looper caterpillar)

Cockchafer grub

Helopeltis punches

Jassid (Green fly)

Jassid infested leaves

Larva of Ascotis sp. (Looper caterpillar)

Larva of Biston suppressaria (Looper caterpillar)

Larva of Cleora sp. (Looper caterpillar)

Larva of Ectropis sp (Looper caterpillars)

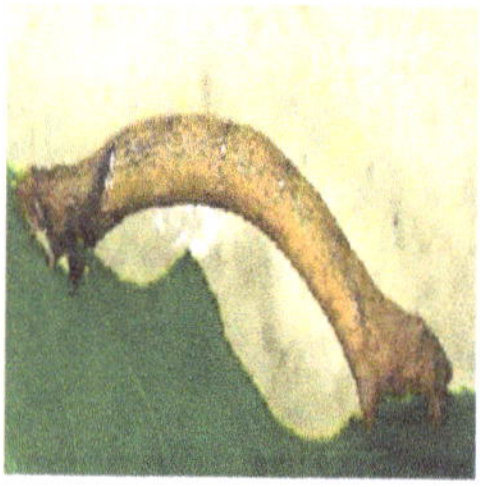

Larva of Hyposidra infixaria (Looper caterpillar)

Larva of Hyposidra talaca (Looper caterpillar)

Moth of Ascotis sp. (Looper caterpillars)

Moth of Cleora sp. (Looper caterpillar)

Moth of Ectropis sp (Looper caterpillar)

Moth of Hyposidra infixaria (Looper caterpillar)

Moth of Hyposidra Talaca (Looper caterpillar)

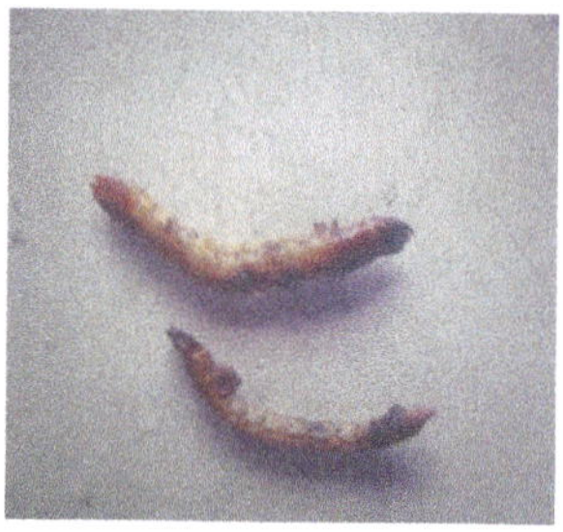

Red Borer

Sand papery lines created by thrips infestation

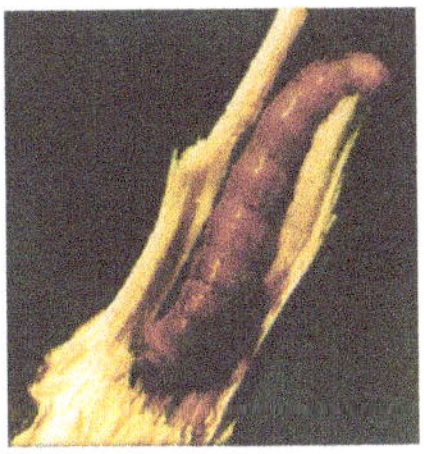

Stem Borer

Tea Mosquito Bug

Termites

Thrips

X- Thrips

PICTURES OF MITES

Pink Mite

Purple Mite

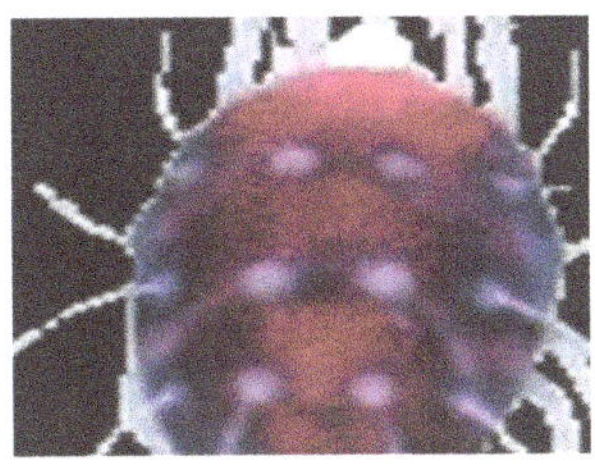

Red Spider Mite

Red Spider Mites on Leaf Surface

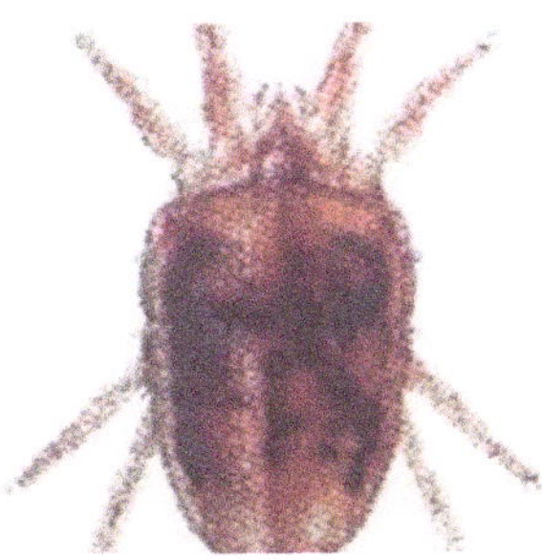

Scarlet Mite

PICTURES OF SHADE TREES AND HEDGE PLANTS

Hedge - Adhatoda vasica

Hedge - Duranta plumeiri

Shade - A twig of Albizzia lebbek

Shade - Albizzia chinensis syn stipulata

Shade - Albizzia lebbek pod

Shade - Albizzia lebbek

Shade - Albizzia Odoratissima Plants in the Nursery

Shade - Albizzia Odoratissima Pod

Shade - Albizzia Odoratissima

Shade - Derris Robusta

Shade - Indigofera Teysmanii

PICTURES OF TEA VARIETIES

Assam Variety – *Camellia assamica* (Masters) Wight.

Cambod Variety – *Camellia assamica ssp Lasiocalyx* (Planchon ex Watt) Wight.

Anthocyanin Pigmentation in Cambod Variety

PICTURES OF VEGETATIVE PROPAGATION

Clonal Cuttings on the Nursery Bed

Netlon Shade

Arched Transparent Polythene Tent

PICTURES OF DISEASE

Black Rot

Blister Blight

Brown Blight

Brown Root Rot

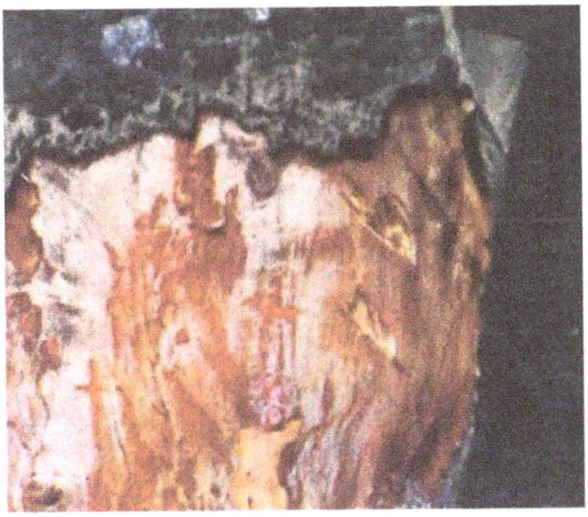

Charcoal Stump Rot

Levelling off Skiff

Leaf Red Rust

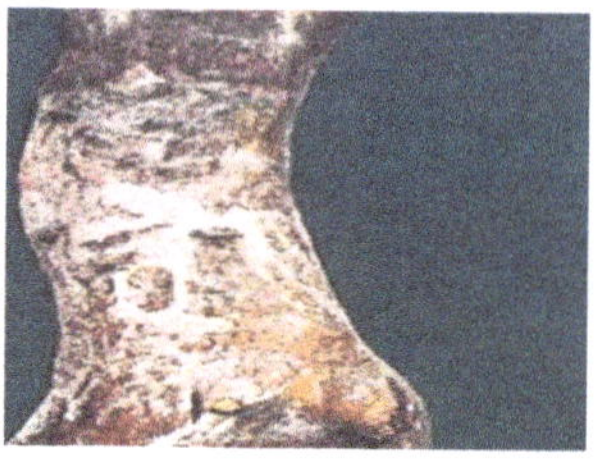

Red Root Rot

Stem Red Rust

Violet Root Rot

PICTURES OF EARTHWORMS

Amyanthas Diffringens

Eudrilus Eugeniae

Eusenia Foetida

Perionyx Excavatus

PICTURES OF PRUNING & PLUCKING

A pruned Tea Section

Medium Pruning Big Cut Marks with Bitumen Paint

Deep Skiffed Tea Section

Medium Skiffed Tea Section

Light Skiff

Levelling off Skiff

Breathers

First FFP

Final FFP

A Pruning Knife

Secateur

Lime Wash of Tea Bushes

A Pruning Machine

A Plucking Machine

9 789359 196268

Printed at Repro India Ltd.